Sanibel & Captiva: A Guide to the Islands
First Edition

ISBN 0-9709596-1-3
ISSN 1536-8947
Library of Congress Control Number:
2001119648

Published by
COCONUT PRESS INC.
5429 Shearwater Drive
Sanibel, FL 33957-2327

Although the Publisher and Authors have made every effort to ensure the information in this publication was correct at the time of going to press, they do not assume and hereby disclaim any liability to any party for any loss or damage caused by errors, omissions, misleading information, or any potential travel disruption due to labor or financial difficulty, whether such errors or omissions result from negligence, accident or any other cause. Readers may call our attention to errors or omissions by writing us at the address above.

Find us online at www.coconutpress.com

Printed in Canada.

Photographs and illustrations
Unless otherwise indicated all photos by Mike Neal, copyright © 2002 Coconut Press Inc. The Publisher gratefully acknowledges the following institutions for the use of their photographs and illustrations: the Bailey-Matthews Shell Museum, the City of Sanibel Historical Committee, the Florida State Archives, the J.N. "Ding" Darling Foundation, the Lee Island Coast Visitor & Convention Bureau, MeriStar Hotels & Resorts, and the U.S. Fish and Wildlife Service. Fish illustrations by Diane Rome Peebles, provided by the Florida Fish and Wildlife Conservation Commission, Division of Marine Fisheries. All maps © 2002 Coconut Press Inc.

Other acknowledgments
We thank the following people for their help in supplying materials and content: Bill Strange and Gary Greenplate at Sanibel Seashell Industries for loaning us shell specimens; Dick Walsh and Tom and Mary Ann Gilhooley for giving us access to their McSpoil scrapbook and sharing their stories; Lauren Davies, Lisa Williams, Karen Aulino and all the members of Sanibel Brownie Girl Scout Troop 641 for its Write Away Try-Its; Patricia Allen and the staff of the Sanibel Public Library for their graciousness and access to their archives; Joseph Pacheco for his poetry; Bird Westall for his essay and insight; Kevin Pierce and Jean Baer for their recipes; Brad Sitton for his golf course tips and tour; Milbrey Rushworth, Sharon Arnold, Kevin Godsea, Leslie Sheffield, Monica Hardy, and Rob Kramer for their hard work gathering photographs and illustrations; and the Conchologists Organization of America for its "You Know You're a Sheller When…" and "Top 10 Reasons to Marry a Shell Collector."

We also thank all the islanders who agreed to be interviewed for this book, including Steve Alexander, Steve Alvarez, Kristie Anders, Francis Bailey, Sam Bailey, John Barden, J.D. Bolden, Gates Castle, Luc Century, Melissa Congress, Jorge Coppen, Terri Cummins, Dave Defonzo, Rob and Cathy DeGennaro, Daniel and Monica Dix, Jim Dowling, Richard Finkel, Liz Fowler, Goz Gosselin, Libby Grimm, Jim Hall, Layne Hamilton, Charlotte Harlow, Malcolm and Susan Harpham, Marty Harrity, Trish Herman, Frank Kik, Dr. José Leal, Erick Lindblad, Dr. Rob Loflin, Jan Manzella, Greg Martinez, Brianne "Bubbles" Meyer, Albert Muench, Dick Muench, Jerry Muench, Ron Orr, Maurice Oshry, Susan Peck, Helene Phillips, Cindy Pierce, Anita Pinder, Bev Postmus, Toni Primeaux, Bob Radigan, Dawn and Joe Ramsey, Andrew Reding, Bruce Rogers, Joan Simonds, Hollie Smith, Robbie Smith, Charlie Sobczak, Jim Sprankle, Sandra Stilwell, Louise Taylor, Nola Theiss, Larry Thompson, Keith Trowbridge, T.C. Tyus, Marcel Ventura, Barbara Von Harten, Steven and Susan Wener, Sandy Zahorchak, Kathy Zocki, as well as all those who participated in our opinion surveys and all the visitors we spoke with on the beaches and in the refuge.

For Micaela, our island girl

Previous page: Thanksgiving morning on Bowman's Beach, Sanibel
Facing page: Jensen's Marina, Captiva

COVER: JACOBS. BACK COVER AERIAL PHOTO: LEE ISLAND COAST VCB

Sanibel

& Captiva

A Guide to the Islands

Julie and Mike Neal

coconut
press

COCONUT PRESS
Sanibel Florida

About the authors

Julie and Mike Neal live on Sanibel's Clam Bayou with their 7-year-old daughter Micaela. Julie, 42, has had a passion for writing — and wildlife — since she was a young girl in Missouri. She taught herself to type at age 9 by writing animal stories on her stepdad's old typewriter. Julie first visited Florida on her high school senior trip. After seeing the ocean, she swore to her friends that she would someday live in the Sunshine State. Mike, 45, grew up just a few miles from Julie, on a lake. He spent nearly every summer afternoon on a water ski, and developed a lifelong love of the water. He became interested in photography by smuggling a camera into rock concerts. Julie and Mike went to the same high school, but didn't meet until a few months after Julie graduated, at a county fair in the summer of 1977. Together they attended the University of Missouri; Julie as a biology major, Mike studying journalism. Fulfilling Julie's dream, the couple moved to Florida in 1980, where Julie finished her education

Julie and Mike Neal

at Florida State. Later they started a small publishing company. After moving to Sanibel, Julie became a volunteer docent for the J.N. "Ding" Darling National Wildlife Refuge, and Mike served as president of the Children's Education Center preschool. Today both enjoy coaching Micaela's soccer team.

How we wrote this book

We took a year to write this book. At first it was tough to figure out what material to include. What did people want to learn about Sanibel and Captiva? To find out, we went right to the source: island visitors. For days we walked the refuge and the beaches, asking tourists what they would want in an island guidebook, and what questions they had about the islands (the most common: "Where's a good place to eat?" and "How can we live here?"). For the answers, we went to island residents. We surveyed hundreds of islanders — getting their insight, advice and tips — and sat down with dozens more for in-depth interviews. Creating the book was a joint project. Julie did the interviews and research. She wrote the first draft, then Mike edited her copy, then Julie edited and proofed Mike's work (this takes a *strong* marriage). Mike took the photos, created the maps and laid out the pages. Julie did the index.

Publications referenced for this book include: "Seashells of North America," by R. Tucker Abbott; "National Audubon Society Field Guide to Florida," by Peter Alden; "Sanibel's Story: Voices and Images," by Betty Anholt; "The Trolley Guide to Sanibel and Captiva Islands," by Betty Anholt; "Eyewitness Books: Shell," by Alex Arthur; "Florida's Hurricane History," by Jay Barnes; "Pirates & Buried Treasure on Florida Islands," by Jack Beater; "Fly-Fishing the Gulf Coast," by Tom Broderidge; "The Nature of Things on Sanibel," by George R. Campbell; "Sanibel-Captiva Cookbook," by the Children's Education Center; "The Sanibel Report," by John Clark; "The Trees of South Florida," by Frank C. Craighead; "Shorebirds & Seagrapes: The Island Inn, Sanibel, 1895–1995," by Sharon M. Doremus; "The Calusa: Sanibel-Captiva's Native Indians," by Elinore M. Dormer; "The Sea Shell Islands," by Elinore M. Dormer; "Fishing Lines," by Florida Department of Environmental Protection; "The Unknown Story of World Famous Sanibel & Captiva," by Florence Fritz; "Capt. Mike Fuery's New Florida Shelling Guide Featuring Sanibel, Captiva," by Capt. Mike Fuery; "Capt. Mike Fuery's South Florida Bay and Coastal Fishing," by Capt. Mike Fuery; "Florida's Fabulous Mammals," by Dr. Jerry Gingerich; "Nature on Sanibel," by Margaret H. Greenberg; "Idea to Reality: An Informal History of the Bailey-Matthews Shell Museum," by William F. Hallstead; "Exploring Wild South Florida," by Susan D. Jewell; "Peterson Field Guides: Southeastern and Caribbean Seashores," by Eugene H. Kaplan; "The Nature of Florida's Beaches," by Cathie Katz; "The Nature of Florida," by James Kavanagh; "Young Naturalist's Guide to Florida," by Peggy Lantz; "Sanybel Light," by Charles LeBuff; "Florida's Birds," by Herbert W. Kale II and David S. Maehr; "Walking Places in Florida," by Diane Marshall; "The Sabal Palm," by Barbara Oehlbeck; "Bicycling in Florida," by Tom Oswald; "Shorebirds of North America," by Alan Richards; "Favorite Birds of Florida," by Dick Schinkel; "The Sibley Guide to Birds," by David Allen Sibley; "Shipwrecks of Florida," by Steven D. Singer; "Walk in the Wetlands," by Robert Slayton; "Sanibel Island," by Lynn Stone; "Fish & Fishing on Sanibel and Captiva," by Norma Stoppelbein; "Birds of South Florida," by Connie Toops and Willard Dilley; "Seashore Life of Florida and the Caribbean," by Gilbert L. Voss; "The Sarasota, Sanibel Island & Naples Book," by Chelle Koster Walton; "Florida Hurricanes and Tropical Storms," by John M. Williams and Iver W. Duedall; "Florida's Fabulous Reptiles and Amphibians," by Winston Williams; "Florida's Fabulous Seashells," by Winston Williams; "Florida's Fabulous Trees," by Winston Williams; "Florida's Fabulous Waterbirds," by Winston Williams; "Beachcombing on Sanibel," by John Harold Wilson and Brenda Wilson Jerman; and Times of the Islands: The Lee Island Coast Magazine.

About this Book

This book tells you everything you need to know to have a great time on Sanibel and Captiva. No matter what your interests are, from beaches and shelling, to wildlife and birding, to boating and fishing, SANIBEL & CAPTIVA: A GUIDE TO THE ISLANDS will help you have the time of your life.

We live on Sanibel, and love our life here. But over the years we've noticed many people who come here don't really know what to do. They go to the beach, but don't know how to take advantage of it. They want to see an alligator, or a spoonbill, but don't know where to look. They rent a boat, but don't see the dolphins jumping behind them! And they drive right past the best restaurants; the places every islander loves. If they only knew what they were missing!

That's why we wrote this book; so *you* don't miss out. So your days on the islands are as fun and rewarding as ours.

SANIBEL & CAPTIVA: A GUIDE TO THE ISLANDS is written for anyone seeking a terrific time on the islands, from the Northern (or European) vacationer wanting a few days of real-Florida fun and sun to the fellow Floridian looking for an idyllic weekend getaway, to the seasonal homeowner here for months at a time.

Headed to the beach? We'll tell you which beach has the most shells, which has the softest sand, even which has the best (and secret) picnic spots.

Want to go shelling? We'll tell you how to find the best shells, with dozens of close-up photos to help you identify your treasures.

Want to see wildlife? We'll show you the islands' wonderful dolphins, manatees, birds, reptiles and other creatures, and tell you exactly where to go to see them in the wild.

Care to go fishing? Our fishing chapter tells you (and shows you) what you'll catch, how and where to catch it, even how to cook it.

We'll also reveal the islands' hard-to-find spots: the cozy mom-and-pop cafes and the tucked-away gourmet bistros; the peaceful island inns and historic resorts; the hidden bike paths and long-forgotten hiking trails. We'll clue you in to museums that recapture the past as well as tram and boat rides that explore the present.

We'll show you how to spot alligator tracks. We'll tell you how to build the ultimate sand castle. We'll take you with us as we spend an afternoon at the beach and a morning out at sea. We'll recommend the best indoor activities, too, such as visiting the world's only shell museum or taking in an intimate live musical.

Divided into 19 chapters, our book begins with island history. Planning Your Trip and Getting Here tell you what to pack, when to come and what things cost; and how to get here by plane, car and boat. The Boating chapter has tips on exploring inland waters, mangrove jungles and deserted out islands. Our Restaurants chapter reviews 64 places to eat and tells you where locals hang out. Other chapters cover the J.N. "Ding" Darling National Wildlife Refuge, the islands' fascinating trees and plants, and what it's like to live here. The final chapters are guides to shops, accommodations and other resources.

Two maps introduce the islands, highlighting major roads, beach access points and places of interest. Other maps show restaurant locations and recommended bike paths. Sidebars and features throughout the book offer inside tips and intriguing detail.

In short, this book is designed to make your time on Sanibel and Captiva a magical and memorable experience.

— JULIE AND MIKE NEAL

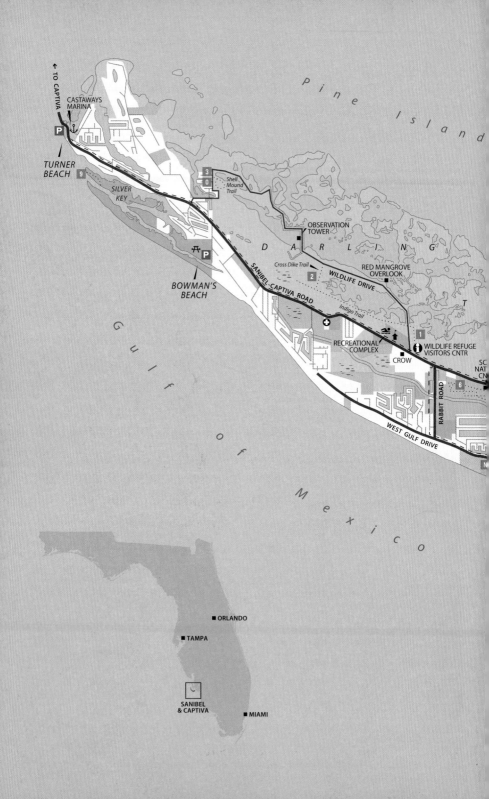

← TO CAPTIVA

CASTAWAYS
MARINA

P

TURNER
BEACH   9

SILVER
KEY

P

BOWMAN'S
BEACH

Pine Island

3
0

Shell
Mound
Trail

OBSERVATION
TOWER

D A R L I N G

Cross Dike Trail   2

RED MANGROVE
OVERLOOK

WILDLIFE DRIVE

SANIBEL-CAPTIVA ROAD

Indigo Trail

T

RECREATIONAL
COMPLEX

CROW

1   WILDLIFE REFUGE
VISITORS CNTR

SC
NAT
CN

6

RABBIT ROAD

WEST GULF DRIVE

Gulf   of   Mexico

ORLANDO

TAMPA

SANIBEL
& CAPTIVA

MIAMI

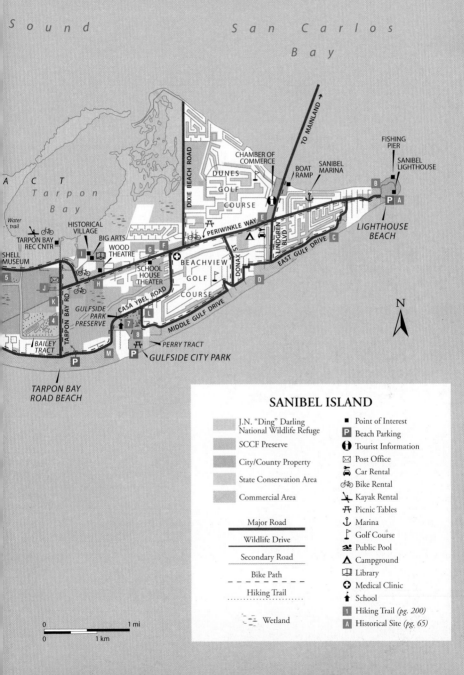

Sound

San Carlos Bay

A C T
Tarpon
Bay

TO MAINLAND

Water
trail

DIXIE BEACH ROAD

CHAMBER OF
COMMERCE

FISHING
PIER

SANIBEL
LIGHTHOUSE

DUNES
GOLF
COURSE

BOAT
RAMP

SANIBEL
MARINA

HISTORICAL
VILLAGE

BIG ARTS

WOOD
THEATRE

PERIWINKLE WAY

E

B

P A

TARPON BAY
REC CNTR

SHELL
MUSEUM

I

G

F

LINDGREN BLVD

LIGHTHOUSE
BEACH

BEACHVIEW
GOLF
COURSE

SCHOOL
HOUSE
THEATER

DONAX ST

EAST GULF DRIVE

C

5

J

H

CASA YBEL ROAD

D

GULFSIDE
PARK
PRESERVE

L

MIDDLE GULF DRIVE

K

4

TARPON BAY RD

7

8

M

BAILEY
TRACT

P

PERRY TRACT

P

GULFSIDE CITY PARK

N

TARPON BAY
ROAD BEACH

SANIBEL ISLAND

J.N. "Ding" Darling
National Wildlife Refuge

SCCF Preserve

City/County Property

State Conservation Area

Commercial Area

Major Road

Wildlife Drive

Secondary Road

Bike Path

Hiking Trail

Wetland

■ Point of Interest
P Beach Parking
Tourist Information
✉ Post Office
Car Rental
Bike Rental
Kayak Rental
Picnic Tables
Marina
Golf Course
Public Pool
Campground
Library
Medical Clinic
School
1 Hiking Trail (pg. 200)
A Historical Site (pg. 65)

0   1 mi
0   1 km

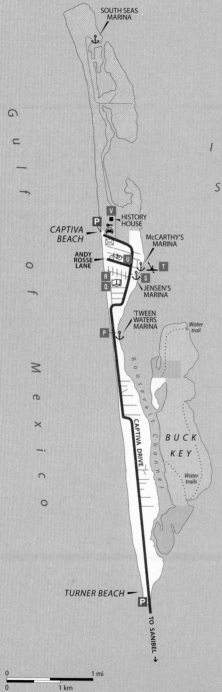

SOUTH SEAS
MARINA

P i n e

I s l a n d

S o u n d

G u l f

o f

V
HISTORY
HOUSE
P
*CAPTIVA
BEACH*
McCARTHY'S
MARINA
ANDY
ROSSE
LANE
U
T
S
R
Q
JENSEN'S
MARINA

TWEEN
WATERS
MARINA
*Water
trail*

P

M e x i c o

Roosevelt Channel

CAPTIVA DRIVE

*B U C K
K E Y*

*Water
trails*

N

TURNER BEACH
P

TO SANIBEL →

| 0 | | 1 mi |
| 0 | | 1 km |

CAPTIVA ISLAND

J.N. "Ding" Darling
National Wildlife Refuge

SCCF Preserve

State Conservation Area

South Seas Resort

Major Road

Secondary Road

■ Point of Interest

P Beach Parking

✉ Post Office

🚗 Car Rental

⚓ Marina

📖 Library

A Historical Site *(pg. 65)*

Contents

Overview

Sanibel and Captiva offer an unbeatable combination: A tropical island wilderness with the convenience of modern life. Only here can you wander a deserted beach, have a gourmet lunch, track down an alligator and see an award-winning play — all in the same day.

Sanibel and Captiva lie off the coast of western Florida, near the city of Ft. Myers. Tampa is 125 miles north; Key West 140 miles south. With more than 16 miles of beaches, a laid-back attitude and great natural beauty and character, the islands are a true 21st century paradise.

As you would expect, the weather is great. Winters are exceptional, with average highs in the mid-70s. Summers are tropical — wet and hot — but balmy Gulf breezes keep things bearable. The sun shines nearly every day, all year round.

As barrier islands, Sanibel and Captiva have open water and sandy beaches on one side; brackish estuaries and mangrove coastlines on the other. Sanibel also has substantial mangrove jungles and unique (for an island) freshwater wetlands.

Sanibel is the larger island, 12 miles long and up to three miles wide — the same size as Manhattan. Captiva, joined to Sanibel's west end by a short bridge, is five miles long and never more than a half-mile wide. The islands are connected to the mainland by the three-mile Sanibel Causeway.

Originally settled by Calusa Indians, Sanibel and Captiva caught the attention of Spanish explorers in the early 1500s. Juan Ponce de Leon named the larger island "Santa Isybella" in honor of Queen Isabella, who funded the New World expeditions. Pirates held female captives on the smaller island, which they came to call Captiva.

American settlers arrived in the 1800s. Farmers grew tomatoes, coconuts and citrus fruit, while beachfront inns attracted wealthy visitors from up north. Hurricanes eventually wiped out farming, but tourism stuck. Early winter visitors included Teddy Roosevelt, Charles and Anne Lindbergh and political cartoonist Jay N. "Ding" Darling, who convinced the federal and state officials to preserve much of the land.

The Sanibel Causeway opened in 1963, and the islands were discovered by the world. At first, rampant development threatened to turn the islands into another Miami Beach — plans were made for 90,000 residents. Sanibel fought back by incorporating in 1974. The new government, all island volunteers, accepted tourism as its future. But it put a near halt to future growth, preserving the remaining undeveloped areas.

Facing page: Sanibel welcome sign

The islands are easy to navigate. Sanibel has just three main roads. Each runs from east to west:

■ Periwinkle Way runs down the middle of the island, from the lighthouse halfway to Captiva. The western half of Periwinkle is Sanibel's main commercial district, home to most of the island's stores and restaurants.

■ Sanibel-Captiva Road (islanders call it "San-Cap") goes out to Captiva. It borders the "Ding" Darling Wildlife Refuge.

■ Gulf Drive runs along the Gulf of Mexico. Technically three roads — East Gulf, Middle Gulf and West Gulf — it's lined with waterfront resorts, inns and timeshares.

Four roads slice across Sanibel north to south: Lindgren Boulevard, Donax Street, Tarpon Bay Road and Rabbit Road.

Captiva has only one road of any significance. San-Cap becomes Captiva Drive when it crosses Blind Pass. This beautiful canopy road runs the length of the island, to the South Seas Resort.

Sanibel was named one of the Top 20 Worldwide Islands by Conde Nast Traveler magazine

Facing page: A Captiva parasailing guide heads off to work

Tranquil beaches

If your ideal vacation includes quiet beaches and warm waters, Sanibel and Captiva are just what you're looking for. The beaches are wild and natural, with more birds than people. The gently sloping sea floor keeps the waves calm and, in some places, lets you wade out hundreds of yards. The crystalline waters stay warm year-round.

Shelling has been the islands' main claim to fame for hundreds of years — ever since Ponce de Leon dubbed the area "la costa de Caracoles," the seashell coast. Sanibel in particular is a sheller's haven. Its east-west orientation puts it in the perfect position, as circulating Gulf currents bring in shells from far offshore. Best of all, the shallow sea floor rolls up many of these treasures unbroken. The awkward, crouching posture of shell seekers even has a name: the Sanibel Stoop.

Thousands of shells wash up with every tide, representing over 160 species. Live shelling is banned, but any visitor is likely to walk upon an abandoned lightning whelk, jewel box or Florida fighting conch, while more serious shunters (SHelling hUNTERS) find rare junonia and golden olives. Beautiful shells can be found on any island beach at any time of day. But you'll find the most at low tide and after a storm. The best spot to look: on a sandbar just offshore at low tide.

Wild encounters

Many visitors to the islands aren't here for the beaches. They come for the wildlife.

The J.N. "Ding" Darling National Wildlife Refuge preserves 6,000 acres of mangrove jungles and virgin wetlands. It's home to dozens of endangered or threatened species. Here tropical and migratory birds roost in the trees, manatees graze in the bays, raccoons amble along the shore, alligators patrol the ponds — all within easy view of people. You can explore the refuge by car, foot, bicycle, kayak or guided tram. A paved road, a water trail and five miles of hiking paths wind through the woodlands and waterways.

Elsewhere on Sanibel, other trails lead you into freshwater habitats. Guided hikes are available at the Sanibel-Captiva Conservation Foundation.

More island fun

Sanibel is a bicyclist's Utopia. Twenty-five miles of paved bike paths run alongside every major road, and occasionally veer off through wilderness areas.

Two unique Sanibel museums are worth a stop. The Sanibel Historical Village and Museum preserves island history with an assortment of buildings from the early 20th century, each fully restored and furnished with authentic artifacts. The Village is passionately maintained by volunteers, who offer knowledgeable, often first-hand accounts of Sanibel life in a bygone era. The only shell museum in the world, the Bailey-Matthews Shell Museum has well organized displays of shells from around the globe, including rare exotic specimens. Exhibits explain the cultural importance of shells through the centuries, from their use in ancient religious ceremonies to their inspiration for architects, painters and sculptors.

And no visit to the islands is complete without getting on the water. Various island outfitters rent power boats, kayaks and Waverunners; parasailing is available on Captiva. Several tour-boat operators offer group trips, including dolphin cruises, natural history tours and sunset voyages. Don't overlook fishing — the islands have dozens of charter captains, experts in finding tarpon, snook, redfish and other rewarding catches.

Charming communities

Sanibel is packed with an enticing charm literally from end to end: from the Victorian-era lighthouse at the eastern tip to the tiny village of Santiva at the west. Life moves at a slower pace on this homespun island. Shop owners open late and close early; the fastest roads have speed limits of 35 mph; night life consists of gathering at

Captiva's bay shore

Island pace: All island roads are 30 mph unless posted. The fastest are only 35 mph. Roads are all two lanes and there are few passing zones. Police write many speeding tickets, particularly along Sanibel-Captiva Road.

At 26 degrees North latitude, the islands are not technically part of the tropics. But Caribbean weather brings a tropical influence to vegetation (i.e., coconut palms) and wildlife (spoonbills).

The area code here is 941. The islands have two exchanges, 472 and 395.

Facing page: The 1884 Sanibel lighthouse

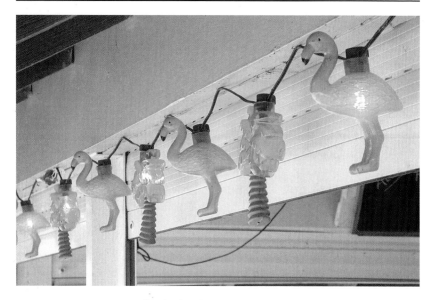

Festive flamingos and palms light up a Sanibel resort beach bar

Sanibel Facts

Area
Total .... 11,600 acres
Preserved .... 8,814 acres
Upland .... 7,930 acres
Mangrove .... 2,800 acres
Elevation
Average .... 4 feet
Maximum .... 13 feet
Rainfall
Yearly .... 42 inches
Season .... June–Nov.
Transportation
Paved roads ...... 60 miles
Bike paths .... 23 miles
Population
Permanent* 5,800
Seasonal peak 33,000
Dwelling units
Single family 3,546
Other 4,317
* includes 6-month residents

Facing page: Captiva shop

the beach to watch the sunset.

Respect for nature outweighs commercial concerns — native flora and fauna are always just a few feet away, even in the largest shopping areas. In fact, Sanibel's character derives in part from what it doesn't have: stop lights, street lights, four-lane highways, billboards, neon signs, fast-food restaurants or other similar man-made intrusions. Many roads and parking lots are paved with shells, not asphalt. Buildings can be no taller than the tallest palm.

On Captiva a recent development boom has bulldozed many small cottages for trophy retreats, but a unique character stubbornly remains. You'll find it in the shops and restaurants along Andy Rosse Lane, along the docks at Jensen's Marina (off Captiva Drive) and at the picturesque Chapel by the Sea (just off Captiva Drive at Chapin Lane). Captiva has no mailboxes; residents pick up their mail at the tiny post office.

Neither Sanibel nor Captiva has a focused downtown. Sanibel's spiritual center is Bailey's General Store. At the western tip of Sanibel is the village of Santiva (SANibel and capTIVA, get it?); it looks like a 1960s vacation postcard. Captiva's heart is the Chapel by the Sea, and it too has a separate village: the South Seas Resort.

Restaurants on the islands include everything from weathered shacks to arty gourmet cafes. The specialty is fresh seafood, especially grouper, crab and shrimp. Locally-owned shops and galleries offer their own brand of charm, with imaginative collections of clothing, shells, art and souvenirs.

Inside Tips for a Terrific Vacation
The best of the islands

Millions of visitors come to Sanibel and Captiva every year. But we're lucky enough to live here. We head to the beach, go for a hike, pedal down a bike path or eat at an island restaurant nearly every day. Over the years we've stumbled upon a few island secrets.

We like to go to the north end of Bowman's Beach at low tide, when the shells can be piled up over 2 feet high. We love to watch the spoonbills come in to roost in the evening across from the refuge observation tower. We return time and time again for the banana pancakes they serve at the Sanibel Cafe, the panini sandwiches at the Bean, and the blackened grouper at Lazy Flamingo. And we never tire of watching the manatees at the South Seas Marina. As far as our family is concerned, the islands are truly heaven on earth.

Each island has a plethora of things to do, but the ones listed here are our best bets: how we spend our time when guests come to visit.

Miles of unspoiled beaches are what most people think of when they think of Sanibel and Captiva. The big debate here: Which beach is the best? Lighthouse Beach, just a short drive from the Sanibel Causeway, is the best bet for day-trippers, and Captiva Beach is where the action is — parasailing trips and Waverunner rentals. But our choice is **Bowman's Beach,** near the western end of Sanibel. The shady park is a great picnic spot, and the beach itself is the epitome of tranquil, natural Florida. Regardless of the beach you choose, **go at low tide.** There's more to do (you can wade far offshore) and more to see — more shells, more wildlife. And don't miss a **beach sunset** — the perfect way to end an island day. The best sunset beaches are those on the islands that face west: Turner and Captiva. The causeway beaches aren't bad, either.

To see the most birds and wildlife, **go to the refuge at low tide.** If you're driving down Wildlife Drive, stop at the first two water impoundments to see scores of herons, egrets and pelicans. Hike down Indigo Trail in the morning to find alligator tracks in the sand. A gang of ibis hangs out where Indigo Trail meets the Cross Dike Trail, back in the marsh on the south side. Whatever you do at the refuge, **spend a lot of time outside your car.**

All island water activities are geared to neophytes — you need no knowledge or experience to rent a powerboat or those zippy little Waverunners, or go kayaking, fishing, sailing or parasailing. Our favorite thing to do on the water is **rent a boat** and head for the out islands of Cayo Costa and Cabbage Key, where you *have* to get a cheeseburger (the best rental boats are at Sweetwater Boat Rentals, 472-6336). Second place: the **Waverunner Safari** at Holiday Water Sports (472-2938 or 472-5111, ext. 3433). It's not cheap — about $100 per person — but skimming and splashing through the open sea, searching for (and finding) dolphins, is a guaranteed thrill. We also find time now and then for a **kayak tour** through Tarpon Bay (at the Tarpon Bay Recreation Center, 472-8900).

Bringing the family? First, **take the tram tour** through the refuge (again from the Tarpon Bay Recreation Center) for a breezy introduction to island vegetation and wildlife; kids will love hearing about the "dinosaur plants" and finding mangrove tree crabs. Then **ride a bike to the hidden Sanibel cemetery.** Finally, make time for a **guided hike** at the Sanibel-Captiva Conservation Foundation (472-2329) — if you can drag the kids away from the touch tank in the visitor's center.

Facing page: An alligator track along Indigo Trail. The bicycle-tire-size tail mark stretches across the sand, with claw marks on either side.

Planning Your Trip

*H*ow do you plan a trip to Sanibel and Captiva? It's easy... just toss a few bathing suits, shorts and T-shirts into a suitcase and head on down. Well, almost. For first-timers, the sheer emptiness of the place can be disconcerting. You sense there's so much to do, but what is it? Neither island is much on commercialism — there are no billboards, neon signs, or airplane banners to give you ideas.

Nonetheless, some people try to wing it. We see them stopped off the side of Sanibel-Captiva Road every morning, appeasing anxious kids in the back seat, poring over a map, debating how to get to Captiva.

To get the most out of an island vacation, you must have a plan. Before you leave your driveway, there are things you can do to make the trip run smoother. First, read this book and familiarize yourself with the descriptions and locations of the beaches and other major areas. Get brochures in advance by calling the Sanibel-Captiva Chamber of Commerce (472-1080). Finally, poll your family and set some priorities — decide, at least in a general sense, what you want to do and, consequently, how much you want to spend. Here are some other tips to help you make the most of your money and time, and have a memorable trip:

What to pack

Island dress is casual, day and night. The easygoing outdoor lifestyle here means **shorts and sandals, T-shirts** and **baseball caps.** Dressing for dinner usually means just changing into clean T-shirts and shorts, even at the most expensive restaurants. Only one restaurant (the Island Inn) requires a jacket for dinner. In the winter add a pair of **lightweight pants** and a **sweater** or light wrap; temperatures can drop into the 50s at night.

Plan on two changes of clothes per day per person; you'll sweat through your clothes within minutes during a peak sun. Hats, shorts and shirts made of 100-percent cotton will keep you

Above: Packed right for the beach
Facing page: Santa keeps watch on Captiva

much cooler than 50-50 polyester blends. Other items to pack include **swimsuits, sunglasses** and **sunscreen.** (Sound like too much? Buy some of it here. Dozens of island shops carry terrific selections of shirts, shorts, swimsuits and accessories.)

You need **cash** to get on the islands: $3 to pay the toll to cross the Sanibel Causeway (the toll booth doesn't take checks or credit cards). Cash is also the only way to pay your admission to Wildlife Drive at the J.N. "Ding" Darling National Wildlife Refuge. **Binoculars** will come in handy to spot wildlife and view the stars. Don't forget your **camera…** or the film, tapes, storage devices, batteries and battery charger to go with it. Other items to consider: **vitamins** and **medications** (including a prescription for more if you think you'll run out); and an extra pair of **contact lenses** or **glasses** in case you lose them.

What you don't need to bring: A suit. A tie. High heels. Makeup. Perfume. Most recreation equipment (golf clubs, tennis rackets, in-line skates, rods and reels) can be rented.

When to go

Any time is a good time to come to the islands. We chat with island visitors daily, and regardless of the time of year they're always having a great time. But each season has its own rewards (and drawbacks):

Spring

Ah! The weather is beautiful, most of the birds are still here, the flowers are blooming. Best of all: the crowds are gone. Early spring is still peak tourist season, and it starts getting hot early (to a Northerner April will feel like August). But most seasonal residents leave at the end of March, and Easter is traditionally the last day of the tourist season. Hotel rates are often discounted after Easter.

Summer

The islands display their true laid-back nature in June, July and August. Crowds are light, traffic no problem. Most summer tourists are other Floridians; others are from Europe (mainly England and Germany). Hotels and resort rates are reasonable. It rains nearly every afternoon for an hour or so, then clears back up. It's buggy too; mosquitoes and no-see-ums are common in the early evening. Nights are warm, and days are hot by 9 a.m. But temperatures are milder than most of southern Florida, as the surrounding Gulf and bay waters cool the island air and bring a soft breeze.

Average temperatures (degrees Fahrenheit)

MONTH	HIGH	LOW
January	74.4	53.2
February	75.5	54.1
March	80.0	58.7
April	84.6	62.0
May	88.8	67.6
June	90.6	72.9
July	91.4	74.6
August	91.4	74.8
September	90.0	74.2
October	86.0	68.5
November	80.7	60.9
December	76.0	55.0
Yearly	84.1	64.7

Gulf water temperatures

Spring / Summer	84.1
Fall / Winter	70.8
Yearly	77.5

Average rainfall (inches)

January	1.84
February	2.23
March	3.07
April	1.06
May	3.87
June	9.52
July	8.26
August	9.66
September	7.82
October	2.94
November	1.57
December	1.53
Yearly	53.37

Facing page: Island philosophy at the Lazy Flamingo restaurant

Cash advances. The following Sanibel banks offer cash advances on Mastercard and VISA:

- Colonial Bank
  520 Tarpon Bay Rd.
- Bank of the Islands
  1699 Periwinkle Way
- Bank of America
  1037 Periwinkle Way,
  2450 Periwinkle Way
- First Union
  2407 Palm Ridge Rd.
- SunTrust of Lee County
  2408 Periwinkle Way

Liquor laws. You must be 21 years of age to drink alcoholic beverages in Florida. Liquor can be purchased on any day of the week. Florida has some of the toughest laws in the U.S. against drunk driving. Don't drink if you're planning to drive.

Business hours. Most island banks are open 9 a.m. to 4 p.m. Monday through Friday; and sometimes Saturday morning. Business offices are usually open 9 a.m. to 5 p.m. weekdays. Stores and restaurants play by their own rules, often open as long as anyone's there. More than once we've called a restaurant to see if it's still open and been asked, "How soon can you get here?"

Facing page: 4th of July fireworks over the Gulf at the South Seas Resort

Fall

"The best time to come here is September, October and November," says Marcel Ventura, owner of Captiva's YOLO Watersports. "The weather is beautiful, and there are no crowds." He's got a point. In September the islands are so quiet some restaurants close. There are no lines, no traffic, hardly any people. Hotel rates are reasonable, and the bugs aren't so bad, either. Even early December's great; the real crowds don't arrive until after Christmas.

Winter

Afternoon traffic in January, February and March can be miserable, restaurant lines are long, hotel rates are high. But this is the weather people daydream about when they are shoveling snow in Ohio or Minnesota. Daytime highs are in the 70s and 80s, it's always sunny and there's no rain. Birding and shelling are at their peaks. And there are more festivals, shows, nature tours and other activities.

Money

What things cost

Item	U.S.$
Shuttle for 4 from SW Florida Intl. Airport	$37.00
Double room at Sanibel Inn, expensive	
Winter/peak	$285 to $309
Summer/value	$159 to $235
Double room at Brennen's Tarpon Tale Inn, moderate	
Winter/peak	$119 to $210
Summer/value	$85 to $169
Double room at Anchorage Inn, inexpensive	
Winter/peak	$89 to $140
Summer/value	$59 to $89
Dinner entree	
At Lazy Flamingo	$7 to $15
At Traders	$19 to $30
Restaurant beverages	
Bottle of beer	$1.25 to $4.25
Soft drink	$1.25 to $2.50
Coffee	$0.95 to $2.50
Admission fees	
"Ding" Darling Wildlife Drive (1 car)	$5.00
Old Schoolhouse Theater	$25.00
Bailey-Matthews Shell Museum (adult)	$5.00
Bailey-Matthews Shell Museum (child)	$3.00
Roll of Kodak film, 36 exposures, at Eckerd	$7.39
Local telephone call (pay phone)	$0.35

Other expenses

Florida has a 6 percent **sales tax** that is added to most items, although not groceries or medical services. Plan to **tip** at restaurants (the standard restaurant tip is 15 percent, but always check the bill first. Some restaurants discretely add a tip to your bill automatically, then still provide a line on the ticket for you to tip again!). Also tip on successful fishing charters. If you're **renting a car,** ask about sales tax, airport surcharges and other fees while booking your reservation. At the counter, be leery of optional add-ons. You probably don't need the optional insurance (rental cars are probably covered in your auto policy you already have; check ahead of time). Beware the gasoline schemes, too. It's almost always the best choice to simply bring the car back with a full tank of gas.

Getting cash

The best way to get cash on the islands is by using your credit or debit card at an ATM. Check your daily withdrawal limit and your credit limits before leaving home.

Suggested itinerary

If you can, allow at least four days on the islands. To stay fresh, work in a combination of active and passive activities every day. Plan a morning bike ride, followed by an afternoon at the beach, or hike the refuge and then spend the afternoon at your hotel pool. You can work in the man-made attractions, such as the Shell Museum and Sanibel Historical Village and Museum, here and there, whenever you need a break from the heat.

Want to experience a true island-style getaway? Here is a suggested three- to seven-day itinerary that does just that, with a sampling of the best the islands have to offer:

The first three days

Day 1: Hit the beach, ride a bike. Get to the beach early. Watch for dolphins. Build a sand castle. For lunch go to Cheeburger Cheeburger; drink a (real) cherry or vanilla coke. In the afternoon, hop on a bike and explore the shops along Periwinkle Way, then head out to the Gulfside Park Preserve off Casa Ybel Road. After a dinner of blackened grouper and cold beer at the Periwinkle Lazy Flamingo, go to the Old Schoolhouse Theater for a fun evening of lightweight music and comedy.

Day 2: Explore the refuge. Get up and out early (by 9 a.m., latest), while the crowds are thin and the weather is

Most island hurricanes have hit between August and November. Meteorologists know days in advance when a storm is brewing and can often tell how strong it will be. Consider postponing your trip if forecasters are predicting a strike — visitors are not allowed on the islands during a hurricane warning. Most airlines will let you move your reservation at the last minute with no penalty; island resorts won't charge you for days you're not here. If you are already here when a storm approaches, your hotel or resort will be able to help you prepare. Follow National Weather Service advisories, television directives and common sense to act responsibly.

Don't wear perfume outdoors on the islands; it attracts the no-see-ums

Serenity now. The best planning advice: follow island time — take it easy.

gorgeous. Stop at the Center for Education at the J.N. "Ding" Darling National Wildlife Refuge and ask the volunteers for information and a map. Watch the 15-minute video to orient yourself, then cruise down Wildlife Drive, stopping often to get out and explore. Have a late lunch at the nearby Island House on Rabbit Road, then visit the Bailey-Matthews Shell Museum, a mile to the east on Sanibel-Captiva Road. Later, have an elegant dinner at Sunset Grill in Santiva, and stroll along the adjacent Turner Beach in the moonlight.

Day 3: Become Jimmy Buffett. Arrive as early as they'll let you at the 'Tween Waters boat dock, rent a boat and head off into the bay. Make your first stop Cayo Costa; anchor on the southern tip and hike around to the Gulf for shelling and exploring. Navigate to Cabbage Key for cheeseburgers for lunch. Write a note on a dollar bill and attach it to the ceiling. Explore the island; climb the water tower. Leave and admire Useppa Island from afar, then roam the bay waters all the way back to the Sanibel lighthouse. Back on Captiva, have dinner at R.C. Otter's (or, for fatter wallets, Keylime Bistro), and reflect on your heavenly day as you listen to island musicians.

Additional days

Add-on Day 4. Do nothing. Go to Bowman's Beach. Fall asleep. Later, float on a raft. Wander around aimlessly. Waste the whole day.

Add-on Day 5: Kayaking and island history. Start with bagels and coffee at the Bean on Periwinkle Way. Then rent a kayak at Tarpon Bay Recreation Center to explore the Commodore Canoe Trail. Munch a hot dog at Schnapper's Hots for lunch. In the afternoon, see island history at the Sanibel Historical Village and Museum. Have a long dinner at Katie Gardenia's, with Katie Kake.

Add-on Day 6: Offshore thrills. Start off early with pancakes at the Lighthouse Cafe on Periwinkle Way (ask for a booth). Then get out on the water. Choose (a) a charter fishing trip, (b) parasailing at YOLO Watersports or (c) a Waverunner tour at South Seas. Stop by Turner Beach for the sunset, then enjoy a fine dinner and a bottle of wine at the Mad Hatter.

Add-on Day 7. Wind down. Get the banana pancakes at the Sanibel Cafe for breakfast. Then drive out to SCCF for a guided

nature tour. Have lunch at Trader's on Periwinkle Way, then take a surrey to Lighthouse
Beach. Check out the nature trails, meander
over to the fishing pier. Look for wentletraps
in the sand. Head back to your hotel for a
shower, then have a pasta dinner at Matzaluna
followed by a play at the Wood Theatre.

*Note: Before you make your plan, get a tide chart
and see when low tide falls (chart times are for
the lighthouse; low tide at the refuge or
Bowman's Beach will be an hour to 90 minutes
later). Adjust your itinerary so you go to the
beach and refuge at low tide (this makes all the
difference; you'll see far more of everything).*

Survival basics

Sunscreen

On Sanibel and Captiva you are about as close to the sun
as you can get in the United States. A good sunscreen is a
must. Look for a lotion with an SPF ("sun protection fac-
tor") rating of at least 30. The real mark of a good sunscreen is quality ingredients: look for Parsol 1789 (or
avobenzone), titanium dioxide or zinc oxide (or Z-Cote).
These help block UVA rays (which damage the inner layers of your skin) as well as UVB rays (which cause sun-
burn). Put on your sunscreen about 30 minutes before
you'll be in the sun, so it has time to penetrate your skin.
Bring it with you, too; you'll want to reapply it often,
even if it claims to be waterproof. If you do get burned,
use an aloe-based after-sun product to ease the pain and
help repair your skin. *(Take the sun seriously. We see many
hot-pink visitors every afternoon, blissfully unaware of the
painful night — and days — ahead of them.)*

**The Sanibel-Captiva
Chamber of Commerce**
(just off the causeway at
1159 Causeway Rd.,
Sanibel; 472-1080; shown
above) has volunteers
(below) ready to answer
any question or guide you
to where you want to go.
Resort, restaurant and
activity pamphlets line
the walls. Ask for a free
flyer showing current
island activities; it's
updated weekly. Don't be
shy; over 200,000 visitors
stop here each year.

Bug repellent

Mosquitoes and no-see-ums (nearly invisible sand fleas that seem drawn to tourist
skin) are a major nuisance here, especially
in the hotter months in the early evening.
Mosquitoes are particularly bad a day or two
after a summer thunderstorm. Carry a good
bug repellent with you and use it liberally.
We use Avon's Skin-So-Soft. *Note: bug re-
pellents that contain DEET can destroy ny-
lon, a common fiber in shorts, shoes and socks.*

Getting Here

I'm finally getting to Florida!" says the visitor. "I'm finally getting *out* of Florida!" says the islander. They're describing the same thing — the wonderful feeling they get when they cross the Sanibel Causeway and head to Sanibel and Captiva.

It's a feeling that says you're going someplace special, that it's time to relax. People say crossing the causeway makes their blood pressure drop 20 points. "It's like shedding your skin," one islander says. "You feel brand new."

Even after many crossings, it's hard to resist glancing at the windsurfers skimming along the waves. Or watching the pelicans gliding alongside your window. Or looking for dolphins out in the bay.

Over a million visitors come to the islands each year. But smart planning has kept traffic and growth under control, and preserved the islands' natural beauty and appeal. No wonder so many other Floridians come here for their vacations, too.

By air

If you're flying to the islands, the view out your window can get you into an island mood even before you land.

Flights from Atlanta and other points north often fly directly over the west coast of Florida, including Sanibel and Captiva. Passengers on the right side of the plane have a terrific view. You'll first see a hodgepodge of small barrier islands, including Gasparilla, Cayo Costa and North Captiva. The water varies from dark blue out at sea to emerald and aquamarine in the bays. Sometimes you can see the sandy bottom underneath.

Eventually Captiva, then Sanibel will come into view. You can identify Captiva by the cleared north tip, where the South Seas Resort's Land's End Village sits. Sanibel is next. It's long, fat and curved like a shrimp. At first glance many passengers don't realize they're looking at Sanibel — forgetting for a moment that most of the island is preserved in its natural state. There are few landmarks, except the three-mile-long Sanibel Causeway and the Sanibel lighthouse, both at the eastern tip.

Southwest Florida International Airport (RSW) *(16000 Chamberlain Parkway, Ft. Myers; 768-1000)* is the area's commercial airport. It's 19 miles east of Sanibel, one mile east of Interstate 75 at Exit 21 (Daniels Parkway). An average of 200 flights take off and land here a day, carrying 4.8 million passengers a year.

February, March and April are the peak months, as Northerners pour into the region to escape the sleet and snow. Many airlines increase their number of flights for the winter season, starting each December. International and domestic charter flights also

---

Facing page: The Sanibel lighthouse has guided the way to the islands since 1884

**Airlines serving
SW Florida International**

- Air Canada
  800-776-3000
- AirTran
  800-247-8726
- Air Transat
  800-470-1011
- America West
  800-235-9192
- American/Am. Eagle
  800-433-7300
- American Trans Air
  800-225-2995
- Balair-CTA
  800-322-5247
- Canada 3000
  877-359-2263
- Cape Air
  800-352-0714
- Condor German
  800-524-6975
- Continental
  800-525-0280
- Delta
  800-221-1212
- Delta Express
  800-325-5205
- JetBlue
  800-538-2538
- LTU International
  800-888-0200
- Midwest Express
  800-452-2022
- Northwest/KLM
  800-225-2525
- Pro Air
  800-939-9551
- Spirit
  800-772-7117
- Sun Country
  800-359-5786
- TWA
  800-221-2000
- United/United Express
  800-241-6522
- US Airways
  800-428-4322

fly a regular winter schedule. The airport's two busiest weeks are Christmas and Easter.

In the terminal you'll find an ATM machine, mail box, duty-free shop and currency exchange. There are a number of fast-food walk-ups and a Chili's Too restaurant.

Airlines

Twenty-five airlines serve Southwest Florida International. Delta is the major carrier. International carriers include American, Continental, Delta, LTU, Northwest/ KLM and US Airways.

Delta has the most daily flights, including eight from Atlanta and six from Orlando. Delta also offers service through its commuter airline, Delta Express (including new-for-2002 service to Newark), and has a working relationship with Comair and Sabena. American and its commuter airline, American Eagle, offer six daily flights from Miami. Continental and the Continental Connection have six flights a day from Tampa.

Airlines continue to expand their service here as more tourists discover the region. American, American Trans Air, Midwest Express, US Airways and Lynx Air International have all recently beefed up service. For the 2002 season Delta Express added a fourth daily flight to Boston, while JetBlue Airways added a second nonstop flight to New York's JFK airport.

Renting a car

You'll find six car-rental companies with check-in counters inside the airport. Enterprise and Thrifty don't have counters at the airport, but have shuttles to their facilities just outside the gates. Always call ahead to make reservations. Southwest Florida is a popular tourist destination, so cars are often sold out, especially in season. You can also rent a car on the islands. The Amoco ServiceCenter *(1015 Periwinkle Way, Sanibel; 472-2125)* rents Hertz cars. There's a Budget office at the South Seas Resort *(5611 South Seas Plantation Rd., Captiva; 472-9600)*.

Taxis and limos

Outside the terminal you'll usually find quite a few taxis waiting. But you don't just hail a cab here. Instead, make your arrangements at the ground transportation booth, located between the terminal and parking lot. Taxis have a fixed-fee structure for service to the islands: $37 for the eastern half of Sanibel (as far west as Tarpon Bay Road); $44 for west Sanibel, including West Gulf Drive;

and $56 to take you to Captiva.

Shuttle and limousine services require advance reservations. You'll pay about the same as a taxi, but the driver will be waiting for you and the vehicles are usually larger and more comfortable.

Most shuttle and limo drivers meet their passengers on the ground floor in the baggage claim area. If you arrive in Terminal A, the waiting area is down the escalator and to your right, in front of Baggage Belt Number 1. If you arrive in Terminal B, the area is down the escalator and to your left, in front of Baggage Belt Number 6.

Some services have their drivers meet you outside the terminal beside their vehicle, on the commercial curb across the street. If you're going through customs, meet your driver in the lobby of the International Arrivals building. If you can't find your driver, pick up a white courtesy phone or stop by the ground transportation booth.

Shuttles and limos serving the islands include:

■ Apple Taxi and Limousine Service *(482-1200 or 800-852-7027)*. A popular van-based service, Apple's rates (for up to three people) are $35 for eastern Sanibel (to Tarpon Bay Road), $44 for western Sanibel, and $54 for Captiva.

■ Sanibel-Captiva Airport Shuttle *(473-0007)*. Another popular service that uses vans and limousines (not stretch). Rates for up to 2 people are $35 to Sanibel, $55 to Captiva.

■ Admiralty/Sanibel Limousine Transportation. *(433-5265 or 800-525-5733)*.

■ Sanibel Island Taxi *(472-4160)*.

Page Field

The area's commercial airport until the early 1980s, **Page Field** *(501 Danley Dr., Ft. Myers; 936-1443)* today caters to charters and private planes, including jets down to two-seaters. It has four runways — the longest is 6,400 feet — as well as a full maintenance facility, fuel and avionics. Budget and Enterprise Rent-A-Car have locations here.

By car

People come to Sanibel and Captiva not only for their natural beauty, but also because the islands are so close to Florida's major cities and other vacation spots. Most out-of-state vacationers who drive here spend the night before in Orlando. Thirty percent of visitors are fellow Floridians, most often from Tampa, Ft. Lauderdale or Miami.

No matter where you're driving from, your first step is getting to Interstate 75, and then to Exit 21:

**Rental cars at
SW Florida International**

■ Alamo
800-GO-ALAMO
Local: 768-2424
■ Avis
800-230-4898
Local: 768-2121
■ Budget
800-227-5945
Local: 768-1500
■ Dollar
800-800-3665
Local: 768-2223
■ Enterprise*
800-736-8222
Local: 561-2227
■ Hertz
800-654-3131
Local: 768-3100
■ National
800-CAR-RENT
Local: 768-2100
■ Thrifty*
800-847-4389
Local: 768-2322

* no counter at airport

Rental car buses are across the street from the main terminal at the Southwest Florida International Airport

Distance to Sanibel and Captiva from other Florida cities (in miles)

Clearwater 145
Daytona 230
Fort Lauderdale 155
Gainesville 250
Jacksonville 305
Key West 290
Miami 158
Ocala 215
Orlando 175
Palm Beach 144
St. Augustine 270
St. Petersburg 130
Sarasota 90
Tallahassee 375
Tampa 145

Sanibel is 18 miles west of Interstate 75, 19 miles west of the airport. That's a 30- to 45-minute drive, depending on traffic.

Opposite page: Relaxed attitude on Captiva

Keep your eyes peeled for "Exit 21," the airport exit. Sanibel and Captiva are the top tourist spots in the region, but I-75 signs make no reference to the islands.

■ **From Orlando or Walt Disney World,** take Interstate 4 south to Tampa, then head south on I-75 to Exit 21. The islands are about 3½ hours from Orlando, 2½ hours from Tampa.

■ **From Atlanta or Tampa,** it's a straight shot down I-75. Driving from Atlanta takes about 10 or 11 hours, not counting stops.

■ **From Fort Lauderdale or Miami,** follow I-75 "north" (you'll actually be heading due west, but the signs will say "I-75 North") out of Fort Lauderdale and across the state, then up past Naples to Exit 21. (I-75 is a toll road from Fort Lauderdale to Naples.)

If you fly in, and rent a car at the Southwest Florida International Airport, follow the signs to exit the airport and head west. You'll get on Daniels Parkway automatically, and pass under Interstate 75 about a mile later.

From the interstate Sanibel is 18 miles away, a 30- to 45-minute drive depending on traffic. Captiva is the same direction — the only road is the one through Sanibel. As you drive you'll see many real estate "available" signs that hint at development to come in this area (so enjoy looking at those Florida cows grazing along the roadside while you can).

Here's how you get to Sanibel from I-75:

■ **Go west on Daniels Parkway** 2.6 miles to Six Mile Cypress Parkway, also called the Ben C. Pratt Parkway. You'll start to see a few directional signs for the islands that use a brown triangle to represent Sanibel and a brown circle for Captiva. Except for these, few signs, either highway or commercial, will mention the islands until you get to the water's edge.

■ **Go left (south) on Six Mile Cypress Parkway** 4.3 miles to Summerlin Road. Six Mile Cypress takes you around the south end of the Fort Myers metro area. On your right you'll pass the Lee County Sports Complex, the spring training home of the Minnesota Twins since 1992. Stop in for a key-lime ice cream cone at Sun Harvest, an orange juice processing company and store 1½ miles down Six Mile Cypress on your left.

■ **Go left (west) on Summerlin Road** 8.0 miles to the Sanibel Causeway Toll Plaza. If you absolutely must eat at McDonald's, your last chance is at the intersection of Summerlin and San Carlos Boulevard, right behind Sharky's Beach Mart. There are no McDonald's, or any other drive-

Don't be surprised if you're delayed for a few minutes by the Sanibel drawbridge opening — consider it part of the charm of getting here. The drawbridge (above) opens every 15 minutes (starting at the top of the hour), whenever a boater requests it. The whole process — opening the bridge, waiting for a boat to pass underneath, and closing it back down — takes only about five minutes. The drawbridge is controlled by the bridge operator, who sits in the tiny control house on the bridge itself.

Boats under 26 feet can pass under the closed bridge. But when a taller boat approaches (usually a sailboat with tall masts), the captain calls the bridge operator on a marine radio to request an opening, and then waits until the next scheduled opening time.

The drawbridge opens more often than others on the Okeechobee Waterway, because water turbulence around the

through restaurants, on Sanibel or Captiva, a point of considerable island pride. Dig through your pockets for three dollars, the same toll since the causeway opened back in 1963.

■ **Cross the Sanibel Causeway** 3.0 miles to Sanibel. Or perhaps just sit still for a few minutes, if the Sanibel drawbridge is up. The causeway islands are man-made, built from sand dredged from the bay. More than 3.3 million vehicles crossed the causeway in 2000, double the traffic of 20 years ago.

The pelicans may fly within inches of your car. Hunting for fish in the water alongside the bridges, they often fly at the same height as your vehicle, and get so close you'll swear they're going to fly right into you. Watch closely and you'll see one spot a fish, suddenly dive-bomb straight into the water, and catch it in its pouch. (Turn off your radio and roll down your windows. If a pelican flies close enough, you may be able to hear its wings rustle.)

Once you're on the island, you'll hit a four-way stop at Periwinkle Way. From here any direction heads to the beach, restaurants and accommodations. Captiva is 10 miles to your right.

By sea

Boating to the islands was the only way to get here until 1963, when the Sanibel Causeway was built. It's still the most scenic and peaceful way to arrive.

Coming in from the north, you'll pass Boca Grande — a getaway for the rich and famous of U.S. society from the Colliers and Vanderbilts of yesterday to the Bushes of today. From the south you'll be charting the same course as Ponce de Leon when he used to sail to the islands from Cuba. You'll spot the Sanibel lighthouse at the east tip of the island, a beacon to sailors for more than a hundred years. If your boat or mast is taller than 26 feet, you'll need to radio the Sanibel drawbridge to open to let you through. It rises on demand at 15- or 30-minute intervals from 6 a.m. to 10 p.m., and anytime with an hour's notice overnight. The water can be turbulent around the bridge.

Except for the dredged Intercoastal Waterway, the waters in San Carlos Bay and Pine Island Sound are shallow — often only a few inches deep, even at high tide. Weekends are crowded with jet skis, flats fishermen, rental boats and more, especially during the winter.

Marinas

Three marinas on Sanibel and Captiva allow overnight docking. All monitor radio channel 16.

Located near the east end of Sanibel, just east of the causeway, the **Sanibel Marina** *(472-2723)* is tucked in just off the bay, between ICW Mile Markers 2 and 3 (6-foot low-tide depth), on a canal. It takes boats up to 110 feet. The marina has fuel and pump-out, parts and service, showers and restaurant, overnight dock and shore power; bait and fishing equipment; a ramp; boat rentals and charters. Hours are 7 a.m. to 5 p.m. daily. Rates are $1.50 per foot. The manager is Stephanie Peterson.

The **South Seas Resort Bayside Marina** *(472-5111, ext. 3447 or 888-777-3625)* was recently named one of the Top 10 U.S. Yachting Destinations by Boating World magazine. Located on the bay side of the northern tip of Captiva, it has 3,600 feet of berthing space and takes boats up to 120 feet. Boaters are allowed to use the resort's barbecue area and gas grills, can take advantage of the resort's recreation program, and can stay in one of the surrounding villas or homes. South Seas has fuel and pump-out (no charge); showers, a restaurant and lodging; overnight dock, shore power, water and cable TV; bait and fishing equipment; a ramp; boat rentals and charters. The marina is open from 8 a.m. to 5:30 p.m. daily. Rates are $2.20 per foot from Thanksgiving through April 30; $1.60 per foot the rest of the year. The dockmaster is John Findley. To get to South Seas from the ICW, head west at Mile Marker 39 (watch for the sign), and proceed 1½ miles west through a privately maintained channel (6-foot low-tide depth). From the Gulf, head through Redfish Pass, which is marked with two green and one red right return floating markers (8-foot low-tide depth).

Part of the 'Tween Waters Inn complex, the **'Tween Waters Marina** *(472-5161)* sits midway down Captiva, on Roosevelt Channel just off the bay. It offers fuel and pump-out; showers, a restaurant and lodging; overnight dock and shore power; bait and fishing equipment; a ramp; boat rentals and charters. Rates are $2 per foot from December 24 through April 30 and holidays, and $1.60 per foot the rest of the year. Hotel guests pay $1 per foot. From ICW Mile Marker 38, take compass heading 215 one mile to Roosevelt Channel, Marker 2. Follow to Green 19.

channel makes it difficult for boats to wait.

The bridge sometimes gets stuck, backing up traffic for hours. Still using technology from the 1960s, it can get locked in the up position, or more commonly not close all the way. It then takes up to two hours to close by hand, with a crank.

The control house is manned from 6 a.m. to 10 p.m. Few boats come through late at night. When they do, a toll-booth operator walks out to the bridge and opens it for them. In the 1960s the bridge was raised every night and kept up — like the drawbridge of a castle — lowering each morning at dawn.

Today the drawbridge is also used to fight crime. Twice in the last few years it has been raised to keep a suspect on Sanibel.

'Tween Waters Marina

History

*F*or such peaceful islands, Sanibel and Captiva have a wild and woolly past. Half-naked Indians beheaded captives here. Pirates are said to have buried treasure, imprisoned young girls and decapitated a princess. A discredited evangelist started island tourism. But the story of Sanibel and Captiva is also a tale of preservation, of people falling in love with these special lands and fighting to protect their physical and aesthetic beauty.

Early history

Sanibel and Captiva are young, only 5,000 years old. They began to form about 3000 B.C., when storms created a sandbar about 6 miles off the coast of today's Florida. Soon mangrove trees began to grow, holding the sand in place to create a permanent island. This became Captiva. Sand continued to build up to the south, eventually forming Sanibel. The whole process took 4,500 years — the lighthouse end of Sanibel didn't exist until about 1500.

Calusa (pronounced "ca-LOO-sah") Indians had lived in the area the whole time. By 1500 there were 40 Calusa villages in Southwest Florida, totaling about 3,000 people. They moved onto Sanibel and Captiva as the islands were being formed. They used the abundance of shells to make middens, or shell mounds, to create everything from homes to burial sites. Using tools such as the knife at right, they ate fish and seafood and built 30-man canoes. For long trips (as far away as Mexico) they strapped two canoes together to make a catamaran.

The Calusa were a tall, proud, handsome people. They wore short deerskin wraps around their waists, nothing else. Disciplined and religious, the Calusa had a stable, advanced society. At least until the Spanish came.

Spanish explorers

Juan Ponce de Leon landed on Sanibel in 1513. Having sailed around Florida from St. Augustine, he stopped here to clean the bottoms of his ships (such as the one shown at right). The explorer named the island "Santa Isybella," after the recently deceased Queen Isabella (popular with explorers, she had helped Columbus just 20 years earlier). "Santa" is Spanish for "saint."

Ponce de Leon attempted to settle the Sanibel area when he returned in 1521, accompanied by 200 people,

---

Left: Visitors arrive at the Sanibel steamer dock, 1908

Oh my God! The Calusa's religious practices could be downright barbaric. As part of the tribe's annual harvest festival, all "nonbelievers" in the village, including shipwrecked Christians, would be sacrificed.

First, the Calusa would cut off the victims' heads, then pluck out their eyes. Believing the eyes contained the soul, the Calusa would offer them to their Harvest God to eat. (The Calusa worshipped three gods. A second controlled the weather; a third, war.)

The tribe found many reasons to sacrifice someone. When a chief died, his servants would be sacrificed so they could continue to serve him in the spirit world. When a chief's child died, other children would be sacrificed, too — so the young boy or girl would continue to have someone to play with.

All Calusa were required to believe in all aspects of the official religion. To make sure they did, each summer the tribe's shaman

50 horses and numerous beasts of burden. It would have been the first European settlement in America. But the Calusa attacked, wounding the famed explorer with a poisoned arrow in his thigh. He sailed to Cuba and died.

The ships of Hernando DeSoto, complete with 640 men, landed on Sanibel in 1539, but were run off the next day. Explorers such as Pánfilo de Narváez and Alvar Nuñez Cabeza de Vaca (1528), Pedro Menendez de Aviles (1566) and Juan Rodrigues (1688) also landed here. They were all, as Rodrigues later wrote, "politely refused."

The Calusa held firm for 200 years, refusing to let the Spanish get a foothold. But their days were numbered. Many died from European diseases such as yellow fever, tuberculosis and measles. Spanish slave ships, armed with the latest guns and cannons, captured hundreds of Calusa in the early 1700s. Outmatched, the remaining natives left the islands. Some moved to Cuba, but most joined the Seminoles, an inland tribe of Creek descendants and escaped slaves. By 1750, the Calusa tribe was gone.

Spanish and British control

The islands were under the Spanish flag until 1763, when Spain traded Florida to England to regain control of Cuba (which it had lost in the French and Indian War). Hoping to appease its disgruntled American colonists, the British promoted Florida as a place to relocate. "Live in cheap Florida, with an estate of your own," one flyer read. Then came the American Revolution.

Pirates and buried treasure

As Americans fought the British up north, Sanibel and Captiva began to be taken over by pirates.

José Gaspar, also known as Gasparilla, was headquartered at Boca Grande on today's Gasparilla Island, four miles north of Captiva. Originally a Spanish naval officer, he stole a boat from his navy and began a new life plundering merchant ships in the Gulf of Mexico.

Gaspar buried his stolen treasures, mainly gold and silver, on Sanibel. He is said to have buried 13 casks and chests of treasure here, worth $60 million. His men, who numbered in the hundreds, buried their smaller caches in the area, too.

In 1801, Gaspar raided the ship of the Spanish princess Maria Louise, traveling across the Gulf from Mexico. On board were 11 Mexican girls, traveling to Spain to go to school. Keeping Maria Louise for himself, Gaspar took the girls to Captiva and locked them away.

About the same time, a slave ship wrecked off the coast

of Miami. One survivor, Henry Caesar, soon became a protégé of that area's reigning pirate, the legendary Blackbeard. Renaming himself Black Caesar, he organized a group of English and Cuban thieves into his own gang. Off Cuba they hijacked a Spanish galleon loaded with 26 tons of silver.

Black Caesar ruled the sea without mercy. Legends say he even burned out the eyes of one of his prisoners, a Baltimore preacher. He moved his hideout, and his silver, to Sanibel when Europeans began to settle in Miami.

The two pirates clashed in 1815. Black Caesar raided Gaspar's Captiva prison, rescuing (or, perhaps, abducting) two of the girls. But before Gaspar could retaliate, the preacher's widow showed up. Trailing him from Miami, she reported Black Caesar's whereabouts to Spanish authorities. They captured the pirate and took him to Key West where he was tied to a tree and burned. The widow lit the match.

Meanwhile, Gaspar had kept Princess Maria Louise locked away on Useppa, a small island between Boca Grande and Captiva. He fell in love with her, but she refused to marry him. Furious, Gaspar chopped off her head. The princess was buried in an unmarked grave on Captiva.

All went well for Gaspar until Spain, which had regained control of Florida after the American Revolution, sold the territory to the U.S. in 1821. Determined to rid its new land of pirates, the U.S. government sent a Navy squadron to capture Gaspar in 1822. The Navy disguised a gunboat as a British merchant ship in Boca Grande Pass, and opened fire on the notorious pirate as he sailed up to overtake them.

But Gaspar refused to be taken alive. As his ship began to sink, he wrapped the anchor chain around his waist, tossed over the anchor, and jumped off the side. His body, his ship, and $1 million of on-board treasure are all supposedly still out there today.

Note that pirate legends are just that — legends. The stories above have been handed down for generations, and are tough to verify. More than a dozen pirate treasure chests have been found along the Florida west coast, but none have been discovered on Sanibel or Captiva.

Settlers arrive

New Yorkers create Sanybel

The town of "Sanybel" was created in 1833 by the New York-based Florida Land Co., making it one of Florida's first "planned communities." Sitting in an office in Man-

(wearing masks such as the ones on the previous page) would run screaming throughout the village, vividly reminding everyone just who was in charge.

The Spanish tried to stop the beheadings. Hearing tales of their shipwrecked countrymen being sacrificed, Spanish priests stopped on the islands many times to attempt to convert the Calusa to Christianity.

But they never could. Jesuit priest Juan Rogel landed here briefly in 1566. Back in Spain, he described the Calusa's religious rituals as "evil beyond belief."

Five Franciscan friars made one last attempt to convert the Calusa in 1697. Taking matters into their own hands, they traveled by themselves to the islands. They landed at night, and cautiously marched into the Calusa village holding candles and crosses. Surprisingly... nothing happened. The Calusa had been scared by the strange sight and had hid in the jungle.

The friars thought they were on to something and marched again the next night. But this time the Calusa were waiting. They attacked the "invaders" and ripped off the Franciscans' clothes. The friars ran naked back to the beach and scurried onto their boats, never to return.

The Jolly Roger. Pirate flags were once common on the seas. But most were not simply black with a skull and crossbones. Each pirate had his own distinct flag.

The purpose of the flag was to scare a ship into surrendering without a fight. So the scarier the design, the better. Most used a skull and bones, but some had a sword, dagger or hourglass (symbolizing that once you saw it flying from an approaching ship, your time was running out). Sometimes the skull had horns, or blood dripping from its mouth.

The first pirate flags were solid red, indicating that the crew would give no quarter (i.e., everyone would die). The name Jolly Roger comes from the French "jolie rouge," meaning "lovely red." Later, a solid black flag meant a ship was prepared for battle. Some pirates would fly a black flag into battle, then lower it and raise a red one if the other ship resisted. Flags with skulls became popular in the early 1700s, and were used by the pirates on Sanibel and Captiva.

hattan, land speculators drew up a village at the east end, complete with a town square, parks and 50 homesites. They divided the rest of the island into 50 parallel tracts of land (often mainly swamp), each running from the Gulf to the bay.

For $500 you got one of each — a lot in town and a farm in the country. "On Sanybel the labors of the husbandman will be amply repaid," the advertisements read, "as the island will become the garden of Florida."

Sanybel barely got started. A few farmers moved in, planting sugar cane, sisal, hemp and pineapple. But the U.S. evacuated the island just two years later, in 1835. It was launching the War of Indian Removal, and afraid the farmers might aid the island's few remaining Indians.

Abandoned, the town washed away later that same year, when a hurricane swamped the island with saltwater.

The Cuban blockade

After the outbreak of the Civil War in 1861, the Confederate government required that all Florida cattle farmers sell their cows to the Rebel Army. It paid $8 to $10 a head, in Confederate dollars. Trouble was, Cuba was already buying Florida cattle. And it paid $30 a head — in Spanish gold.

Farmers from Tampa to Miami continued to sell to Cuba, bringing their cattle to the stockyard at Punta Rassa (right), which held thousands of cows.

The Confederate government responded by building a naval station on Sanibel at Point Ybel (where the Sanibel lighthouse stands today), complete with sloops and gun boats. It created a blockade, preventing any Cuban boats from entering or leaving the area.

Magic beans

In 1862, during the Civil War, the U.S. passed the Homestead Act. It allowed any U.S. resident (i.e., Northerner) to claim up to 160 acres of selected land, free of charge, as long as he or she farmed it for five years. When he heard the news, Union soldier William Allen got an idea. Stationed on Key West, he had noticed how effectively Army doctors were using castor oil to treat yellow fever. Once the war was over, Allen thought, he could file a homesteading claim in Florida, grow castor beans, and get rich.

When the war ended, Allen found his spot: Sanibel.

Not that far from his army contacts at Key West, the island's east end had just been opened for homesteading.

In 1866, Allen and his brother set up a castor-bean farm on the same ground as today's Brennen's Tarpon Tale Inn. Within weeks their dream started coming true — the beans were growing like crazy. In 1871 they received the title to the land. They were set for life.

Or so it seemed. On Oct. 6, 1873, a hurricane covered Sanibel with 5 feet of water. Their crops and equipment ruined, the Allens had no choice but to abandon their salt-soaked homestead.

Their castor beans, ironically, are still here. Some seeds survived the storm, and today the beans grow wild.

Cuban settlements

Cubans set up fish ranches — small villages to salt mullet and prepare it for shipping — on the islands in 1871. One was near the spot of today's Sea Horse Shops on Periwinkle Way. Each consisted of a main building to prepare the fish, another to hold ice and other supplies, and a group of thatched cottages and sheds.

The buildings were made without nails, "Gilligan's Island" style: just palmetto fronds wrapped around red mangrove frames. Captains went back to Cuba in the summer, but the guides and workers stayed on the islands all year. Up to 150 Cuban fishermen lived on Sanibel by 1877.

Sanibel lights up

After the Civil War, Cuba continued to buy Florida cattle. San Carlos Bay and the port of Punta Rassa were getting busier every day. But many steamers came in at night, and navigating the notoriously shallow water was a risky proposition. After the 21-ton Sea Bird wrecked in the bay on Aug. 29, 1880, the U.S. decided to aid the merchants by erecting a navigation lighthouse on Sanibel. The design called for an open frame, like that of an oil derrick and similar to others in South Florida and the Keys, so strong winds could blow right through it.

The lighthouse was lit on Aug. 20, 1884. Federal officials marked the occasion by opening nearly the entire island to homesteading. (The east end, including land as far west as today's Bailey Road, was kept in government hands for decades.) With no rights to the land, the Cubans deserted their fish camps. (But they continued to fish the area for 75 years, until Fidel Castro took power in the late 1950s.)

The first American settler was William Reed, who moved here with his son from Maine in 1887. Frank Bailey

Captiva's 10-year-old landowner. One of the youngest landowners in Florida history was 10-year-old Ann Brainerd. From Canada, she had moved with her family to Buck Key, a small bay island next to Captiva, in 1895. In 1901 she met William Binder, Captiva's largest landowner. She told him how this one piece of his land, less than an acre, was the prettiest spot she had ever seen. Charmed, Binder said he'd sell it to her in exchange for a small gold coin she was carrying. The young girl agreed.

Tragically, young Ann stepped on a rusty nail just a few days later, and died from tetanus. Her family buried her on her land, which over time became the Captiva cemetery. Today islanders still visit the girl's grave (below), placing seashells on the tombstone in her honor. Her family is buried next to her. The grave of William Binder is a few steps away.

SANIBEL HISTORICAL COMMITTEE

Lighting the Way

The Sanibel lighthouse is the first along the Florida west coast as you sail north from Cuba. It was built to help Cuban cattle steamers find San Carlos Bay and the port of Punta Rassa at night. The 104-foot-tall structure has an open frame with an iron skeleton. It stands on four legs braced by girders and tie bars. An early example of prefab construction, the lighthouse was made in New Jersey in large sections, then shipped to Sanibel on a schooner. But the ship struck a shoal two miles offshore and broke apart. The lighthouse fell overboard. Except for two small fittings, all the parts were recovered by Key West salvagers and used for construction. The light was lit on Aug. 20, 1884.

The first light burned coal. Each evening the keeper carried the coal, as well as a five-gallon can of kerosene, inside the narrow metal tube and up the 127 spiral steps. At sundown he climbed back up to light the burner with a match; at dawn he climbed up again and blew it out. The light could be seen 16 miles out at sea.

The lighthouse was converted to electricity in 1962. A year later a technical glitch caused it to go dark for a week, ruining a 79-year perfect record. An early lighthouse lens is now on display at the Sanibel Historical Village.

Trade with Cuba ended with Fidel Castro's rise to power, and Punta Rassa's days as a port ended when the Sanibel Causeway was built in 1963.

But the lighthouse remained. When the Coast Guard decided it was no longer needed in 1972, islanders convinced the Guard to change its mind, and two years later got the lighthouse declared a national landmark.

Obsolete or not, the lighthouse still operates. Boat captains still use it, but its most important function is symbolic — as a reminder of Sanibel's unique history and spirit.

SANIBEL HISTORICAL COMMITTEE

Charles Henry Williams was Sanibel's lighthouse keeper from 1910 to 1923 and his family homesteaded a year later. By 1889 100 Americans were on the island — 40 families living in just 21 homes. (The population would stay under 300 for 65 years, until the first small subdivision was built in 1954.)

Captiva was settled by accident — literally. Austrian shipwreck victim William Binder washed up on a Captiva beach in 1885. He spent 10 days on the island recovering from his wounds, and fell in love with the place. Legends say he then swam to the mainland and immediately filed a homesteading claim.

Paradise on earth

In the late 1800s Sanibel was beautiful, even prettier than today. With no Australian pines or Brazilian pepper, the island was like a large subtropical park, with long vistas, open glades and exquisite color. The interior was a natural prairie, carpeted with long grass and wildflowers, dotted with clumps of sabal palms. The weather was great, the beaches terrific, the wildlife incredible.

As Northerners would soon discover, it was an ideal place for a vacation.

The Promised Land

Island tourism got its start when an unorthodox traveling evangelist moved to Sanibel in 1889. From Kentucky, Rev. George Barnes stumbled upon Sanibel by accident, while in nearby Punta Gorda visiting friends.

Taking him out for a tour of San Carlos Bay, Barnes' friends didn't realize the tide was going out. The boat ran aground a few hundred yards off Sanibel. Stuck in the sand until high tide, Barnes and his friends decided to wade over to the island and look around.

As he explored Sanibel, Barnes fell in love with its beauty. Soon he became convinced that God had planned the whole boating mishap for his benefit, that Sanibel was his own personal Promised Land. Barnes returned to Kentucky, packed his bags, and moved to the island with his family to begin a new life.

In Kentucky, Barnes had been tried for heresy for his nontraditional beliefs, which combined Christianity with mysticism from India. Disgusted with his fellow Presbyterians, he decided to establish a nondenominational island church. To attract passing sailors, he built it just a few hundred feet

'Remember the Maine!' Working for the International Ocean Telegraph Co., George Schultz ran Sanibel's cable relay hut. It connected an ocean-floor telegraph cable from Key West with another to Punta Rassa. On Feb. 15, 1898, he received a message from Key West. It read "At 9:40 Tuesday the U.S. Battleship Maine was blown up in Havana Harbor." Schultz relayed it to the mainland, which sent it to Washington. That incident started the Spanish-American War.

Today, the site of the cable hut is marked by a small plaque. Look for it just off the road, just before the lighthouse.

The Church of the Four Gospels was built by Rev. George Barnes in 1889

FLORIDA STATE ARCHIVES

from the shore on his Gulfside homestead.

Calling it the Church of the Four Gospels, the optimistic Barnes put in enough pews to seat 300 people — three times the island population.

Well-known in the South and East as the "Mountain Evangelist," Rev. Barnes continued to travel, preaching to thousands at each stop. He loved to tell others of Sanibel, and worked tales of the island into his sermons.

Fish tales

Still with plenty of room on his property, Barnes and his son, William, next built a 30-room inn. Barnes named the building "The Sisters" because his unmarried daughters had their rooms there.

Sanibel became world-famous a few years later, on March 12, 1895. New York tourist W.H. Wood caught a tarpon in San Carlos Bay with a rod and reel — a feat thought to be impossible. Until then, fishermen had only caught tarpon with clumsy shark hooks and chain lines, or sometimes with harpoons. The accomplishment meant even novice anglers could now catch this trophy fish. It attracted the attention of the entire sporting world, and many national magazines carried the news.

Rev. Barnes added the tale to his traveling sermons.

Barnes' believers

"It's paradise on earth!" Rev. Barnes proclaimed at a revival meeting in Kentucky in the summer of 1895. In the

Hooking tourists. Early settlers caught tarpon with clumsy shark hooks and chain lines (above). But tourists soon found it possible to use simple rods and reels (below).

FLORIDA STATE ARCHIVES

Facing page: Rev. George Barnes and his wife pose on their Sanibel beach

Sanibel's Casa Ybel inn was a true family affair. May Barnes (above) and her sister lived in the inn, while brother William (below, right) managed the front desk. "Casa Ybel has a home-like atmosphere with a spirit of camaraderie," he wrote. "It's a clean, wholesome place, with no bar, no gambling, no so-called 'high living,' but rather sane, pleasant enjoyment amid congenial surroundings." Rates were $2 per day, $12 per week.

crowd were Will Matthews and his wife, Harriet. Will, an accountant, had always dreamed of becoming a farmer.

To Will Matthews, the Rev. Barnes' tale was spellbinding — Sanibel sounded like the perfect spot to fulfill his farming dreams. So he packed up his family and headed south. He and Harriet moved in at The Sisters before the year was out, and soon purchased an island farm.

But once they bought a farm, the Matthews nearly bought *the* farm. Having no experience growing crops of any kind, Will watched his dream quickly turn to dirt (or, in this case, sand). To make a living the couple began to take on boarders at their home, giving Rev. Barnes a little friendly competition.

For decades, Rev. Barnes and the Matthews offered the only accommodations on the island. Barnes renamed his inn Casa Ybel in 1903. The Matthews daughter, Charlotta, named their place the Island Inn in 1936. Both resorts are still operating today.

A Victorian vacationland

Within no time Sanibel became a popular destination for wealthy Easterners. Businessmen came to escape the tensions of the industrial age. Sportsmen came for the game. The elderly came for the warm salty air.

But more than anything, guests came to relax.

A 1904 Casa Ybel brochure explained the island's unique appeal:

"Here one may be thoroughly rested. With plenty to amuse him, without turning a hand to amuse himself, one may loaf

to his heart's, or body's, content. To watch the bathers; to stroll idly on the beach; to sit on the pier and watch the fishers haul in their prizes — these be recreations which re-create.

"The somnolent breezes rustling the palms and the steady swishing monotone from the beach will make him sleepy long before he is tired. He will retire at an unconscionably early hour to a night of dreamless slumber. It is the almost universal testimony of island visitors that they 'sleep like a log' on Sanibel. Many go to bed soon after dinner, because they can't keep awake."

Families loved it here. Children could play safely outdoors all winter long. The beaches made clean playgrounds, and, unlike the beaches of the Florida east coast, these had no undertow, giant waves or sudden changes of depth.

Fish, fowl and frills

Staying for weeks at a time, nearly all guests went fishing, with guides arranged by the inns. "On Sanibel the angler can revel in piscatorial abandon and cover himself with fish scales," wrote "Camping and Cruising in Florida" in 1902.

Shelling became one of Sanibel's greatest attractions, not just to the women and children, but to nearly all the men as well. Initially scoffing at the mania for picking up "worthless shells," a man would typically

Just down from Casa Ybel, the Matthews' home had grown into a full-fledged inn by 1910, including the beachfront "Barracks" building above

A couple shows off their finds at the 1911 Casa Ybel Shell Day

A Victorian Holiday

Sanibel becomes a vacation paradise, 1899–1911

Below: A couple arrives on the island after the six-hour cruise from Punta Gorda, circa 1910. This boat, the Pastime, was 11 feet wide and 57 feet long, similar to those used today to transport tourists within Walt Disney World. **Bottom:** Beachcombers at Casa Ybel display the latest fashions, February 1911. PHOTOS: SANIBEL HISTORICAL COMMITTEE

Left: A father compares the height of his son to that of a shark he caught, circa 1908. **Bottom:** A manta ray caught at the Casa Ybel dock, 1911. Still common in the Gulf, manta rays are not dangerous to man.

Top: Left to right: Janet Butterfield, U.S. Olympic marksman Charles Tayntor Sr. and his sons Harold and Charles Jr., circa 1905. Tayntor Sr. practiced his shooting skills on the Sanibel beaches. **Above:** Fishing off the pier at Casa Ybel, circa 1909.

Left: Heading out for a duck hunt from Casa Ybel, February, 1908. **Below:** Casa Ybel owner Rev. George Barnes (left) and friends catch sheepshead and mullet for the hotel dinner, 1899. The man in the middle is holding a grain, a spear with barbed forked tips used to impale fish as they swim by.

Traveling to Sanibel at the turn of the century was an adventure in itself. First, you took a train to Lakeland, Fla., then caught a second, notorious train (above) down the west coast (travelers would often complain about the drunken, pistol-carrying locals onboard). At Punta Gorda you hopped on a steamboat for a six-hour trip to Sanibel. Finally, you climbed onto a mule-drawn wagon for the 40-minute ride to your inn.

Casa Ybel hunters with turkey, deer and, at far right, a gator belly skin

come back from his first beach trip with both pockets bulging and both hands full. Guests displayed their best finds at the annual Casa Ybel Shell Day.

Other organized activities included tennis, croquet and dancing. At Casa Ybel, a corner of the reading room was reserved for the "flirtatiously inclined."

The inns took their guests on mule-drawn wagon rides along the beach. Longer trips went to the lighthouse or Blind Pass, with oysters served out of the shell for lunch. Some visitors shipped down horseless carriages and drove them along the shore.

Hunting was popular. Instead of using blinds and decoys, guests would simply walk down the beach, as hundreds of birds would bob up and down in the surf. If a hunter got within range before they flew, he (or she) took a shot. Inland, hunters would approach ducks on their hands and knees, and crawl within range.

Some guests took day trips to Punta Rassa to hunt quail, turkey and deer. For fun, some guests shot sandpipers on the beach, and occasionally even large wading birds.

This being the Victorian era, a proper sense of decorum was always important. Women wore formal dresses at all times, unless swimming at the beach. Dinners at the inns were coat-and-tie affairs, with no alcohol served.

The food was terrific. A typical dinner featured an oyster appetizer; followed by fresh fish caught by the guests themselves that afternoon; complemented with tomatoes, peppers and other vegetables picked that day by island farmers; topped off with homemade key lime pie.

FLORIDA STATE ARCHIVES

Farming

Fields of gold

Homesteading on Sanibel in the 1880s, farmers grew tomatoes and limes. They also found success with eggplant, cucumbers, watermelon, radishes, peppers and bananas. They had a nine-month growing season, October to June.

But it wasn't easy. During the summer the hot sun would burn up the fields. Mosquitoes, gnats and sand flies ("no-see-ums") made farmhands miserable. Drinking water smelled and tasted like sulfur.

A deep freeze hit the island in 1898, complete with a few snowflakes — the only snow in Sanibel history. Icicles hung from the palms, and crops were damaged.

But just as the farmers began to have second thoughts, they struck gold. Connecticut businessman Bradley Plant built a railroad and series of luxury hotels down the Florida west coast, and began running freight steamers from the end of the line, at Punta Gorda, down to Sanibel. Now island farmers could sell their crops nationwide.

Each morning a steamer would arrive from Punta Gorda. A commission merchant would hop off, shake hands with the farmers, examine their produce and cut deals. Soon Sanibel tomatoes were everywhere. They were especially popular in New York grocery stores.

Sanibel became an active small town. The Sanibel Baptist Church (now the Colonial Bank) opened in 1909. Kids went to school at the one-room Sanibel School on Periwinkle Way (now the Old Schoolhouse Theater).

Captiva grew, too. A separate community, it built its

Shallow-draft steamboats (including the Uneeda, above, circa 1911) came to Sanibel at the turn of the century, making commercial farming possible. To handle the business, farmers Frank and Harry Bailey took over a wharf warehouse and created the Sanibel Packing Co. (below, circa 1900) and a general store. Kids packed the produce, earning 4 cents per crate, three dollars a day. A hurricane destroyed the wharf in 1926. But Bailey's General Store still exists today.

SANIBEL HISTORICAL COMMITTEE

Captiva's Snyder School drew students such as Bill Kimball (above), later president of the Kimball Piano Co. as well as Sanibel's Island Inn

Sanibel farmer Harry Bailey poses in front of his lucrative key lime groves, circa 1923

own school (now the Chapel by the Sea) and opened the Snyder School, a summer camp for wealthy boys.

A trip to Fort Myers took three days. First you caught the noon steamer, arriving in Fort Myers that night. You conducted your business the next day. Finally, on the third day, you caught the morning steamer back.

A 1910 hurricane wiped out the Barnes church. But farmers replaced it with the nondenominational Sanibel Community Church in 1914 (still in operation). The land was donated, and islanders held food and bake sales to raise money for materials. The Barnes family helped out too, giving the church some pews salvaged from the Church of the Four Gospels. Volunteers built the new church when they could spare time from their fields.

But the thriving towns were not to last. That same year the U.S. entered World War I, and most young island men left to fight. Potash, vital for growing vegetables in sandy soil, was commandeered by the government for military purposes. Meanwhile paved roads had come to the mainland, giving farmers there a huge competitive advantage in getting their crops to market.

Two hurricanes dealt the final blows. Most fields were ruined by a 1921 storm, which covered Sanibel and Captiva with saltwater. The remaining crops were wiped out by the tremendous 1926 hurricane, when a 14-foot storm surge completely swamped the island.

Farming was finished.

Fresh Florida Grown Key Limes
MATURED ON TREES
$1.50 per carton containing about one hundred Limes according to size delivered to you to you by express prepaid.
$2.50 for one quarter box about twice the size of carton prepaid by express to you.
FLORIDA KEY LIMES ARE THE ORIGINAL LIMES AND HAVE THEIR OWN DISTINCTIVE FLAVOR

(Buy American Grown Products)

SANIBEL PACKING COMPANY
Sanibel, Florida

Coconuts and key limes

Now nearly deserted, land on the islands became quite a bargain. The depressed prices caught the attention of Captiva visitor Clarence Chadwick. He had just struck it rich as the inventor of the Checkwriter, an automatic check-writing machine.

Chadwick figured the islands were still fine for farming, as long as he grew salt-tolerant crops. He purchased the north end of Captiva and created a large working plantation that grew coconuts and limes. Soon Chadwick became one of the largest key lime distributors in the world.

But another hurricane hit in 1935. The islands were again covered with water, and most of Chadwick's lime trees were heavily damaged. Noticing how tourism had replaced farming on Sanibel, Chadwick decided to do the same. He closed down his plantation, and reopened it as the South Seas Resort.

Ferries boost tourism

Regular ferry service to Sanibel from Punta Rassa began in 1926. The ferry carried seven cars and a handful of passengers. It made four trips per day. Thomas Edison and Henry Ford were regular passengers. Looking for new rubber sources, they searched for exotic plants to take back to Edison's winter laboratory in Fort Myers.

With hurricanes stirring up the Gulf floor every few years, "rare" shells were everywhere. Visitors would carry flour sacks to the beach and fill them with Florida cones

The 'Shark Factory.' The Ocean Leather Co. operated a shark rendering plant on Sanibel in the 1920s. Located on the bay side of the island near the lighthouse, the "shark factory" processed sharks into lubricants and vitamins (from the livers), fertilizer and soap (from the meat and intestines), glue (from the cartilage) and, yes, leather. At least that's what the plant called the shark skin, once it was tanned. A windmill propelled the stench of this foul-smelling place to farm homes throughout the east end of the island.

The Best ferry loads cars headed for Sanibel. Rates were $1 per vehicle — often just a horse or mule and its cart — and 50 cents per passenger.

SANIBEL HISTORICAL COMMITTEE

SANIBEL HISTORICAL COMMITTEE

Casa Ybel had become a thriving full-service resort by the 1920s. Note the well-groomed grounds.

Prohibition was not enforced on the islands. Stills were common, with moonshine made from sugar cane. Cubans would stop at the wharf to trade rum and acquadente, a potent liquor. Below, lighthouse keeper Roscoe McLane (left) buys liquor from an island bootlegger.

SANIBEL HISTORICAL COMMITTEE

and olives, as hundreds lay on the sandbars. Scientists from Harvard and the Smithsonian Institution visited, gathering shells for study and field-identification books.

More tourists came each year. New resorts and restaurants opened. Charles and Anne Lindbergh honeymooned on Captiva in 1929. Poet Edna St. Vincent Millay vacationed on Sanibel. Plans were made for a causeway to connect Sanibel to Pine Island, already connected to the mainland. Then came the Great Depression.

An isolated world

Tourism nearly disappeared during the Great Depression, and the islands reverted back to their own isolated world. The few remaining residents lived off the land. Many ate alligators, sea turtles and gopher tortoises. Many homes were abandoned, then taken over by poor black families. The Baptist Church was converted into the Sanibel School for Colored Children. To socialize, islanders gathered at the Sanibel Community House. A mix of upper-crust transplanted Yankees and down-home Florida crackers, islanders would schedule Shakespeare readings one night, square dancing the next, then ballroom dancing on the weekend.

But it was all by kerosene lamp. Electrical service wouldn't reach the islands until 1941.

Early developers get Dinged

The state of Florida declared Sanibel, Captiva, and their contiguous water a state wildlife refuge in 1930. But the designation did nothing more than ban hunting. Ten years later developers convinced the state to sell off 2,000 acres on the bay side of Sanibel at 25 cents to $1 an acre.

Jay Darling, visiting from Iowa, found out about the plan. Having recently served as the first head of the U.S. Biological Survey (the forerunner of the U.S. Fish and Wildlife Service), he knew the vital role the islands played as a wildlife habitat. Using his ties in Washington, Darling delayed the sale, and instead arranged for a federal lease of the land in 1944 and its designation as the Sanibel National Wildlife Refuge in 1945.

The developers were blocked. Eventually purchased outright by the federal government, the land formed the beginning of today's J.N. "Ding" Darling National Wildlife Refuge.

Bombs and torpedoes

The island ferries were requisitioned for troop use in World War II, and few visitors came to the islands. But it wasn't quiet here. The west end of Sanibel became an artillery and bombing range. As the U.S. military trained pilots and bombardiers over Bowman's Beach, stray bullets would occasionally rip through roofs and rainwater tanks.

German submarines began showing up off the beaches, spying on the activities and hunting for U.S. supply barges coming from the North. A Coast Guard detachment was stationed at Casa Ybel, and a spotting tower was built next to the lighthouse. Visitors walking along the beach reported seeing periscopes in the water.

One resident witnessed a sub attack. "It was almost dark and I was looking west at the last faint light," she reported. "Suddenly, without any sound, I saw a barge blow up. It burst into flames and floated past the Island Inn. It was still a terrific fire when it reached the lighthouse."

Only 75 people lived on the islands during the war. All were photographed and fingerprinted.

Today artillery shells still occasionally wash up on Bowman's Beach.

J.N. "DING" DARLING FOUNDATION

Pulitizer Prize-winning political cartoonist Jay Norwood Darling signed his cartoons "Ding," a contraction of his last name. Working for the Des Moines Register, he was syndicated in 130 newspapers. His favorite topic was conservation.

The U.S. Coast Guard built a submarine spotting tower next to the lighthouse during World War II

SANIBEL HISTORICAL COMMITTEE

FLORIDA STATE ARCHIVES

Hurricanes

Twelve hurricanes struck Sanibel and Captiva from 1870 to 1960. None have hit since. (Does the word "due" come to mind?)

A hurricane is a large rotating system of wind. It forms out of a mass of thunderstorms in warm water far out at sea. It builds slowly. When the wind begins to rotate in a closed circle, the storm is labeled a tropical depression. When sustained winds reach 39 mph, it's called a tropical storm. When wind speeds reach 74 mph, the storm becomes a hurricane. In the Northern Hemisphere, hurricanes always spin counterclockwise.

A strong hurricane can devastate a coastal community. Winds can blow out windows and then blow off roofs. Typically 6 to 8 inches of rain falls at speeds up to 100 mph. Sometimes the storm's collision with land creates tornadoes and whirling vortices, with their own counterclockwise spinning winds of up to 200 mph. But the real damage comes from the storm surge — a literal rising of the sea whipped up by the strong winds. The water can rise up to 20 feet, and it comes up in just a few hours.

Weather officials began naming hurricanes in 1950. Unless retired, the names are recycled every six years. The names for the 2002 season include Arthur, Bertha, Cristobal and Dolly. In 2003 the first four storms will be Ana, Bill, Claudette and Danny. There are 21 names each year (the letters "Q," "U," "X," "Y," and "Z" are not used).

Hurricane season is officially June through November. But since 1900, every hurricane that has hit the islands has come in September or October.

Here's a recap of those storms:

■ **Oct. 17, 1910.** A large section of land east of the Sanibel lighthouse washed away in this hurricane, and the Church of the Four Gospels at Casa Ybel was destroyed. The barometric pressure dropped to 28.40 inches. A lone white-tailed deer wandered around the lighthouse afterward, the last deer ever seen on the island.

■ **Oct. 25, 1921.** This hurricane broke Captiva into two parts, creating Redfish Pass. Most of the islands were covered with a few inches of saltwater.

■ **Sept. 19, 1926.** This hurricane brought a 14-foot storm surge to Sanibel, ending commercial farming. The storm came across the state from Miami, where it turned that area's real estate boom to bust. At the time it was the most severe hurricane in U.S. history.

■ **Sept. 3, 1935.** It rained for nine straight days before this storm hit. Water was armpit-deep on Bailey Road, waist-deep in the middle of Sanibel. "It was a plain of water stretching the width of the island," recalls Sam Bailey, then 11.

■ **Oct. 18, 1944.** Wind gusts reached 100 mph during this storm, which came straight north from Cuba. A Cuban fishing boat in San Carlos Bay tried to make it to Sanibel for safety, but was blown onto a grassy flat (near where the center causeway island is today). The crew drowned. All the burrowing owls on Sanibel were killed. Quail, once common, disappeared. Two-thirds of the Casa Ybel cottages were tossed off their foundations. A few days after the storm government surveyors accidentally started a large fire on Sanibel, when they lit a palm frond to defend themselves against wasps.

■ **Sept. 17, 1947.** Winds reached 110 mph. Parts of the islands were six feet underwater. The lighthouse keeper said he looked down from his tower and saw no green.

■ **Sept. 10, 1960.** Hurricane Donna was the killer storm people fear. Fortunately it hit in September (when few tourists were on the island), and in 1960, when the islands had less than 600 residents. The most destructive hurricane in Florida's history until Andrew in 1992, Donna spun up the west coast without losing speed. The eye came across at Bonita Beach. Barrier islands from Sanibel to Gasparilla (facing page) were hit hard.

Winds reached 121 mph. The Fort Myers News-Press (below) reported that Australian pines "fell like ten pins" across Periwinkle Way in stacks up to 15 feet high, making it impossible for cars to get to the ferry. Coast Guard helicopters made last-minute evacuations, landing in 60 mph winds. The 6-foot storm surge made officials fear Sanibel would be cut in two: the island had new man-made canals dug throughout the east end.

Here's how one resident described the storm: "Our neighbor's dock crashed into our house. Acting as a battering ram, it was propelled by each successive wave, finally forcing down a wall. It churned inside the children's bedrooms. Then Donna broke in through the roof, through the windows and through the front door, exposing the entire house and furnishings. Our belongings took flight. Some pieces were found around the post office. Others sailed to the lighthouse. Many, of course, were never seen again."

With no way to leave, 38 people crammed into the Bailey home on Periwinkle Way for safety. "Even there," one said later, "the bay quietly crept to the back yard and gradually rose. The palm trees bent double. We mopped and mopped as the rainwater came at the house horizontally. It sneaked under the doors, around the windows, over the sills. There was no gaiety, no laughter, no drinking, no partying. Most of us were terrified."

About half of the homes on Sanibel were damaged. Many had their roofs torn off. Some were without electricity for 26 days. The island did not split in two, but most of the east end was underwater. The next day President Eisenhower declared all of Lee County a disaster area.

FORT MYERS NEWS-PRESS

FEROCIOUS HURRICANE CAUSES HUGE DAMAGE, 3 DEATHS HERE

121-Mile Winds Rake Area Toppling Trees, Buildings; No Grave Injuries Reported

Donna Was Here

Quick Action Planned To Restore Utilities

Huge Truck Jackknifes, Dives in River

SANIBEL PUBLIC LIBRARY

Shipwrecks. Here's a partial list of ships that have wrecked off the waters of Sanibel:

- The William & Frederick was lost off Sanibel on April 2, 1832, sailing from Apalachicola to Key West

- The Union captured the Confederate schooner Ida on March 4, 1863. It was run aground on Sanibel and destroyed.

- Sailing on its maiden voyage from Cedar Key to Charlotte Harbor, the 38-ton steamer Huntress was blown off course during a storm, swamped and sank off the coast of Sanibel on Oct. 6, 1873. It contained $4,500 of building materials.

- Sailing from Key West, the 21-ton Sea Bird foundered just east of Sanibel on Aug. 29, 1880. It was a total loss.

- The 207-ton Martha M. Heath was stranded 2½ miles west of Sanibel on May 5, 1884. Coming from New York, it contained the prefabricated Sanibel lighthouse and coal for its burners. Nearly all of the lighthouse pieces were recovered and used.

- Sailing south from Tampa, the 65-foot, 40-ton freighter Chase was stranded on Sanibel on Jan. 10, 1928

- The 83-ton freight vessel Athenian burned 25 miles off Sanibel on May 26, 1973

Developers move in

After the war the tourists returned. More small inns opened along the beaches. When visitors drove their cars off the ferry, they'd be met by innkeepers, handing out brochures. The charming islands were more popular than ever.

Then mosquito control came to the islands. The Sanibel River was created by the U.S. Army Corps of Engineers in 1950, connecting the island's freshwater swales — a mosquito breeding ground — into one single canal. Trucks began spraying for the pests in 1953.

Now outside developers were again back in force. Having failed to buy up the refuge land a decade earlier, this time they were after everything else.

With little to stop them, they began dredging the wetlands, piling up the spoil to create roads and homesites. The east end of Sanibel was sliced up into canals. A St. Petersburg corporation made plans to fill in Tarpon Bay.

"Ding" Darling died while plans for the Sanibel Causeway were being reviewed. Strong resident opposition didn't stop Lee County, and the causeway was built in 1963. But it didn't go to Pine Island, as once planned. Developer Hugo Lindgren convinced the county to build it straight from the mainland, across the bay, directly into his new Sanibel subdivision. (Incredibly shortsighted, the decision ensured that Fort Myers and Punta Rassa, with the southernmost bay on the Florida west coast and a history of Cuban trade, could never again be major ports.)

The county then rezoned the island for maximum profit potential. In 1967, it announced that all non-refuge areas of Sanibel would be developed. Forget the "heaven on earth" stuff, the county said, high-rise condominiums would be built throughout the island. Plans called for 90,000 residents, a density equal to Miami Beach. And county planners didn't forget the refuge — a four-lane highway would be built right through it, and continue on to Captiva, North Captiva, Cayo Costa and Gasparilla.

With the county in their pocket, developers had a field day. They scraped and burned native vegetation and dredged and filled wetlands. Too impatient for a sewer system, they plopped down septic tanks by each new building. Raw sewage began to leak into the remaining waters.

The natural Sanibel of "Ding" Darling was disappearing. The spiritual Sanibel of Rev. Barnes was being forgotten. (Sold, ignored and falling apart, his Casa Ybel would be torn down in the late 1970s, replaced by all-new buildings.) In their place came a blacktop jungle of subdivisions, strip centers and parking lots, covering more of the island day by day.

But there was still hope. A grassroots conservation movement was gaining strength, and more people were appreciating the islands' natural qualities. Taking over the mission "Ding" Darling had begun, the Sanibel-Captiva Conservation Foundation was firmly established. Most tourists were nature lovers; shelling was more popular than ever, and with the opening of Wildlife Drive visitors were now coming for the birds as well as the beaches.

Sanibel residents and business leaders began to realize they had to take control of their destiny. If they could stop the developers, they could preserve much of Sanibel's character while also securing their tourist economy.

Sanibel takes control

On Nov. 5, 1974, residents voted to incorporate. The island of Sanibel became the city of Sanibel.

The new government — all volunteers — immediately issued a moratorium on building permits. Its goals were to preserve the island and its quality of life. The first mayor was Porter Goss, now a U.S. congressman.

Completed in 1976, the Sanibel Plan limited the island to a maximum of 7,800 dwelling units (later increased to 9,000), and allowed only a small percentage to be built each year. New buildings had to be farther off the beach, and none could be taller than 45 feet. The plan also protected wildlife and native vegetation.

Developers sued. Sanibel prevailed, arguing that an overcrowded island could be a death trap in a hurricane.

The islanders had their island back.

The problems of paradise

"Sanibel shall remain a barrier island sanctuary and a small-town community," says the 1997 Sanibel Vision Statement. "It shall be developed only to the extent to which it retains its quality of sanctuary. It will serve as an attraction only to the extent to which it retains its qualities as a sanctuary and community."

But creating such a Utopia has brought a new threat to the island: gentrification. With property values soaring, the island is changing, not always for the better.

"Before, younger people and people with limited incomes could come to Sanibel, fall in love with it, and move here with their family," Francis Bailey says.

SANIBEL PUBLIC LIBRARY

Tourism was heavily promoted after the causeway connected the islands to the mainland. The islands' Chamber of Commerce published brochures (above, 1965), while the Florida Dept. of Tourism posed models on the beach (below, 1971).

FLORIDA STATE ARCHIVES

Big Mac Attack

Once you get accustomed to the islands' unique beaches, weather and wildlife, you begin to notice something else is different here, too. There's no McDonald's.

The fast-food giant tried to come to Sanibel in 1993. But residents said no. They fought off Big Mac by organizing quickly, getting tons of publicity and finding McDonald's Achilles' heel. Here's how it happened:

McDonald's announced plans to build on Sanibel on Dec. 8, 1992. It paid $571,000 for an acre of land on Periwinkle Way across from Bailey's General Store, where the SunTrust Bank sits today. The news alarmed islanders. Before Christmas a grassroots group, called McSpoil, organized and publicized McDonald's plan.

The island was united in its opposition. Hotel owners didn't want to damage Sanibel's "get-away-from-it-all" atmosphere.

Restaurant owners didn't want the competition. Even kids didn't want Big Mac. In a poll at the island elementary school, 53 percent of the children said they wanted McDonald's to stay away. "Our students feel as though they are the keepers of the island," explained then-principal Barbara Ward.

Tourists were the most vocal opponents. They flooded Sanibel newspapers with letters.

"A true vacation spot is a place you can escape from one's normal lifestyle, a place to take it easy and relax," wrote a visitor from Pittsburgh. "Fast food should remain in the fast lane back home. A McDonald's on Sanibel will change what the island has to offer."

"McDonald's is in airports, the subways, hospitals, college and high-school cafeterias, stadiums. Dear God, please, don't they have enough?" wrote a visitor from Kansas. "Must even Sanibel be covered with cups and cartons, straws, bags and Happy Meals? I can see the osprey nests in Ding Darling now — built from Big Mac cartons and apple pie boxes."

"I am a 14-year-old girl from Atlanta," another tourist wrote. "I have been coming to Sanibel and Captiva with my family for nine years. One of the reasons I love the islands so much is because they are not very commercialized. Living in Atlanta, I eat at McDonald's pretty often. Believe me, it is not that great. In fact, it's not even good."

McDonald's ignored the commotion. "This is a vocal minority that only appears to be a majority," a spokesman said. But McSpoil had a great strategy: Don't try to kill the entire McDonald's monster. Just kill its profit center — the drive-through window. The group petitioned the city of Sanibel to ban drive-through restaurants.

The city commissioned two independent studies; both said allowing drive-throughs would be a mistake. "Sanibel needs to be very particular about the land use mix that it permits," one report said. "The island markets itself as 'unique.' If it has all the normal strip commercial uses, that sense of uniqueness is lost."

Sanibel officials agreed. The drive-through ban became law in August, 1993. McDonald's fought the law for two years. But five days before Christmas, 1995, it withdrew its application. A McDonald's spokesman said, "Public opinion had nothing to do with it."

Sanibel lawmakers went even further in 1996, passing a complete ban on what it called "formula" restaurants — those with three or more locations that have the same or similar name, menu, uniforms and exterior. But since McDonald's pulled out, no other franchised restaurants have even contacted the Sanibel Planning Department. Apparently if you knock down the leader, the whole gang runs away.

"But now many hotels have long minimum stays, so we're getting less young and middle-class people right off the bat. For most who do come to visit, our high housing prices make it too hard to move here."

In peak season some hotels require a weekly stay, with room rates up to $350 a night. A home that sold for $79,500 in 1979 appreciated to $425,000 by 2001. Typical new homes start at $600,000. Often the only people who can afford them are second-home buyers.

A monoculture of wealthy seasonal residents (something no islander wants, not even the snowbirds) is threatening to destroy Sanibel's character. With many homes sitting vacant most of the year, the close-knit island community is starting to suffer. The island's rock-solid support of its churches, schools and environmental organizations is still there, but showing cracks.

"Diversity makes a city strong," Bailey says, "and gentrification works against diversity."

Captiva, which chose not to incorporate, has been transformed. Much of the soul of the island — small bars, restaurants and cottages — has been torn down for elite vacation homes and trophy estates. Just a handful of fulltime residents remain, and there is no longer a central town.

"Sanibel and Captiva are a great experiment," says former mayor Mark "Bird" Westall, now a nature guide. "When you protect the environment and limit human habitation, property values go up. But then people come here not just for the nature, but also because of the exclusivity and value of the place."

Historical sites

Boxed numbers below refer to locations on the Sanibel and Captiva maps on pages 6 through 8.

Sanibel Historical Village & Museum

This unique museum **1** *(850 Dunlop Rd., Sanibel; 472-4648)* is a collection of five furnished buildings from the early days of Sanibel. Included is a small home, the tiny original post office, and the Bailey's General Store of 1935 (a building later used as the island's co-op preschool). The garage behind the store houses Bailey's original 1926 Model-T delivery truck. A small garden has castor beans and key lime trees. Our favorite spot here is the Rutland House, a 1913 home that is a marvel of true island architecture. Authentic furnishings, artifacts and pictures in each building offer a glimpse of this bygone era. Exhibits explain the history of Sanibel.

Dude, where's my grouper? During the 1970s many bales of marijuana floated up on island beaches. Locals called it "square grouper." Teens, and some adults, collected the bootleg bounty as it washed ashore. Though soaked with saltwater, it could be dried out in a microwave and used as a brownie ingredient.

Keeping it real. The city of Sanibel banned neon signs, stop lights and four-lane roads in the 1970s. Islanders feel they don't fit in a true small-town atmosphere. Police direct traffic by hand.

Following page: Authentic buildings at the Sanibel Historical Village & Museum include (top to bottom, left to right): the 1913 Rutland House; the 1898 Burnap Cottage; the 1930s-era Bailey's General Store; the tiny original Sanibel Post Office. Bottom right: a new addition being trucked in.

For a hands-on experience, ask one of the volunteer docents to let you play the Rutland House piano, work the General Store scale, ring an old ship's bell or pull weeds in the garden. During the holidays children leave letters to Santa at the post office.

The museum is only open Wednesday through Saturday, 10 a.m. to 4 p.m. (1 p.m. from June to Aug. 15). The museum is closed on holidays and from Aug. 15 through Nov. 1. A small donation is requested.

The Sanibel lighthouse

You can walk right up to the 1884 Sanibel lighthouse A but the still-operating structure is not open to the public. The 104-foot tower and the adjacent keeper's quarters (where some city employees now live) were placed on the National Register of Historic Places in 1974.

The lighthouse is easy to get to, at the east end of Periwinkle Way. To walk to it, park in the Lighthouse Beach parking lot and walk down the long, shaded boardwalk to your left. To drive or bike up, turn left just before you get to the parking lot.

Calusa shell mound

The Shell Mound Trail O in the J.N. "Ding" Darling National Wildlife Refuge encircles an ancient Calusa Indian shell mound, where the tribe tossed its leftover shells. As you walk along the boardwalk, look off the trail for this huge pile of bleached-white shells. The ⅓-mile walk

You can make pencil rubbings of headstones at the Sanibel (above) or Captiva cemeteries

The Colonial Bank K
(520 Tarpon Bay Rd.; 472-1314) was a Baptist Church during the farm boom. Later, it became a school for black children, then a black church.

Captiva's Island Store *(11500 Andy Rosse Ln.; 472-2374)* was built in 1921. The converted rooming house was once owned by "Ding" Darling.

The Chapel By the Sea *(adjacent to the Captiva cemetery, 11580 Chapin Ln.; 472-1646)* was built in 1901 by Captiva's first settler, William Binder, as a schoolhouse. Church services are still held here.

winds though a shady hammock of subtropical vegetation unique in southern Florida. The Shell Mound Trail is just off Wildlife Drive, near the end of the five-mile road. A clean boardwalk and interpretative signs make the walk easy and worthwhile.

Island cemeteries

Most of the graves at the two tiny island cemeteries are from the Victorian era. Headstones with the inscription "C.S.A." identify Civil War soldiers who served the Confederate States of America. The **Sanibel Cemetery** is off the bike path that runs between Middle Gulf Drive and Algiers Drive. Bordered by a small wooden fence, it has just a handful of graves, including that of an unknown shipwreck victim. The lovely **Captiva Cemetery** is built on an ancient Calusa shell mound, tucked under gumbo limbo and seagrape trees. Visitors often place shells on the grave of Ann Emma Brainerd. Though only 10 when she died, Ann owned the land where the cemetery now rests.

Other places of interest

Ruins of **Sanibel's ferry landing** are still visible west of the island fishing pier. Walk up the beach to see the filled-in small canal and pilings. This was the dropping-off point for island visitors from 1928 until 1963,

What's in a Name?

Blind Pass. The Spanish named the cut between Sanibel and Captiva "Boca Ciego," which can mean either "blind" or "shut up." The pass once snaked blindly through the area, reaching the Gulf on the east side of Bowman's Beach. Today the second meaning makes more sense, as this now straight pass often fills with sand.

Bowman's Beach. This Sanibel beach was named after Robert Bowman, who attempted, and failed, to homestead on Sanibel in the late 1800s.

Buck Key. This small island on the bay side of Captiva was home to many deer before the 1910 hurricane.

Captiva. From the original Spanish name, Isle de los Captivas ("Island of the Female Captives"). Legends say pirates held female captives on the island in the 1700s and early 1800s. Other "captivating" tales say Calusa Indians held Juan Ortíz captive here for 11 years in the 1500s and that the Spanish captured Calusa slaves here.

Casa Ybel. "The House of Sanibel." Islanders pronounce it as one word, "CAHS-ah-bel," keeping the "Y" silent. *See Ybel, below.*

Dixie Beach. Where the ferry Dixie docked in the 1920s, this beach is on the bay side of Sanibel near the lighthouse. Ironically, Dixie Beach Boulevard does not go to Dixie Beach. Lee County officials named the wrong road.

Periwinkle Way. Sanibel's main thoroughfare is named after the flowers the Bailey family planted along the road. County planners wanted to name it Sanibel Boulevard.

Punta Rassa. Where the Sanibel Causeway meets the mainland, this area was once an active port. "Punta Rasa" is Spanish for "flat point." An extra "S" was added by mistake.

Rabbit Road. This mid-Sanibel road connects West Gulf Drive with Sanibel-Captiva Road. Years ago, when it was nothing more than a path through the saw grass, it was thick with marsh rabbits. A small street off Rabbit Road is called Bunny Lane.

Redfish Pass. Formed by the 1921 hurricane, this pass between Captiva and North Captiva had no name for two years. But in 1923 it became so thick with redfish, fishermen said you couldn't put a hook in the water without catching one.

Roosevelt Channel. This channel between Captiva and Buck Key is named after Theodore Roosevelt. He stayed on a houseboat here in 1913, studying devilfish and sharks.

San Carlos Bay. The body of water between Sanibel and the mainland was named after Carlos, the area's Calusa Indian chief. The "San" was added in the 1920s.

Sanibel. A contraction of Santa Isybella ("Saint Isabella"), the name Ponce de Leon gave the island in 1513 in honor of recently deceased Queen Isabella. The Spanish called the island San Ybel ("EE-bell") in the 1700s, often shortening it to simply Ybel. The island was incorporated as "Sanybel" in 1833 and finally spelled "Sanibel" before the turn of the century. Some local historians also point out that the phrase "P. de S. Nibel" (South Level Entrance) appeared next to the island on a Spanish army chart in 1768, referring to the southern entrance to San Carlos Bay. Take off the "P. de," replace the remaining period with an "A," and the phrase becomes "Sanibel."

Snook Motel. This longtime Sanibel inn was not named after the fish. The original owner was Floyd Snook. It's now known as the Waterside Inn.

Useppa. This island north of Captiva was named "Josefa" in the early 1800s by fisherman José Caldez who had a boat of that name. Italian fishermen called it "Guiseppe" (Italian for Joseph) in 1870. American mispronunciation led to "Useppa."

Ybel. The shorthand version of San Ybel, the spelling of the island in the late 1700s. Point Ybel (where the Sanibel lighthouse is), then, means "the point of Sanibel."

The original Bailey house (1896) is still the private home of the family. It sits on Periwinkle Way in front of Donax Street.

How do you say "Sanibel?" Though most everyone pronounces it "san-a-bell," the first homesteaders called the island "san-a-bull." Their descendants still use this pronunciation today.

Old-time native Floridians are often called "crackers." The term refers to the state's early cattlemen, who did their herding with long whips that went "crack!"

For more on island history check out the following books, available at island libraries and most bookstores:

■ "Sanibel's Story," by Betty Anholt

■ "Sanybel Light," by Charles LeBuff

■ "The Nature of Things on Sanibel," by George R. Campbell

■ "The Sea Shell Islands," by Elinore M. Dormer

■ "The Unknown Story of Sanibel and Captiva," by Florence Fritz

■ "The Calusa: Sanibel-Captiva's Native Indians," by Elinore M. Dormer

■ "True Tales of Old Captiva," by various Captiva residents

when the Sanibel Causeway made ferry traffic obsolete.

Two charming 1940s resorts worth a look are the **Seaside Inn** C *(541 East Gulf Dr., Sanibel; 472-1400)* and the **Gulf Breeze Cottages** D *(1081 Shell Basket Ln., Sanibel; off Nerita St., which connects Middle and East Gulf Drs., 472-1626)*. The lobby of a tiny **Bank of America** E branch *(1037 Periwinkle Way, Sanibel; 472-5575)* is decorated with an inspiring collection of historical photographs. Just down the street is the **Sanibel Community Church** F *(1740 Periwinkle Way, Sanibel; 472-2684)*, started in 1917 by English minister George Day (pews from the 1889 Church of the Four Gospels are still used inside); the 1896 **Old Schoolhouse Theater** G *(1095 Periwinkle Way, Sanibel; 472-6862)*, Sanibel's one-room schoolhouse for grades 1 through 8 until 1963; and the 1928 **Sanibel Community House** H *(2173 Periwinkle Way, Sanibel; 472-2155)*, still the island's meeting place today, and home to the Sanibel Shell Fair.

One of Sanibel's oldest surviving homes is the 1891 **George Cooper House** J *(630 Tarpon Bay Rd., Sanibel)*, now part of the Olde Sanibel shops. Sights along West Gulf Drive include **Casa Ybel** M *(2255 West Gulf Dr., Sanibel; 472-3145)*, a 1970s-era resort built on the site of the original 1889 inn; and the **Island Inn** N *(3111 West Gulf Dr., Sanibel; 472-1561)*, still going strong after 100 years, and rich in history in its own right (check out the photo albums in the lobby). Have breakfast or dinner here and go back in time to a softer, slower era. (Dinners still require jackets for men, slacks or dresses for women.)

Captiva, too, has its share of island stories and style. The dining room of the **Old Captiva House** P restaurant *(at 'Tween Waters Inn, 15951 Captiva Dr., Captiva; 472-5161)* is filled with original drawings by J.N. "Ding" Darling. Stop by for an authentic Old Florida meal (try the key lime pie). Teddy Roosevelt, and Charles and Anne Lindbergh, ate here often.

For another dose of Old Florida shake hands with the Jensen brothers at **Jensen's Twin Palms Resort and Marina** S *(15107 Captiva Dr., Captiva; 472-5800)*, a hangout for Captiva characters. Within walking distance is **McCarthy's Marina** T *(15041 Captiva Dr., Captiva; 472-5200)*, another island spot with a colorful past.

The **Captiva History House** V *(at the entrance to South Seas Resort, near Chadwick's restaurant, 472-5111)* is one of the original worker cottages from the resort's days as a key lime plantation. Today it displays old photographs and memorabilia. The museum is open five days a week; admission is free. Historical tours of the resort depart from here Tuesdays and Thursdays at 3:00 p.m.

Island Originals

Sam and Francis Bailey know Sanibel. They were born on the island. Grew up here. And, at 78 and 80 years old, are still here, running Bailey's General Store.

Their father, Frank Bailey, moved here with his parents in 1894. Frank's dad was a retired railroad man, and his mom was sick and needed a warmer climate. Sanibel seemed an ideal spot to relocate, as long as someone in the family could earn a living.

That job fell to Frank. Just 21 years old, he became a farmer. He grew watermelons, carting them down to the steamer landing to ship to Cuba, Key West and New Orleans.

Sam, Francis and John Bailey, circa 1930

Five years later, in 1899, he opened a dockside packing and shipping business, the Sanibel Packing Co. As a sideline, he bought out the island's lone retail store, which he renamed Bailey's General Store. When the farming boom hit the islands that same year, Frank Bailey, then 26, became Sanibel's Main Man — the one person everyone knew.

Sons Francis, John and Sam grew up on an isolated, subtropical island. They had a life few can even dream of. As young boys, they'd go fishing at least once a week. Sam remembers he could catch mullet anywhere — bay, Gulf, inland. Francis remembers a different fishing story. He's fond of the morning when "I caught a needle fish, John caught a big shell with a hole in it, and Sam caught his big toe on a fish hook!"

They'd swim in the Gulf nearly every day. But forget swimsuits — these guys just tossed off their clothes and hopped in. "Only once did some fishermen in a boat happen to see us," Francis remembers, "and one of the fishermen turned out to be a fisher-woman!"

The boys went to the old one-room Sanibel School (now the Old Schoolhouse Theater) through eighth grade. There were 32 students, one teacher. To get away from things, Francis would raise his hand to use the outhouse, then simply stand outside and listen to the meadowlarks sing.

As young adults the Baileys became as well known as their father. In 1951 Sam got the swamp cabbage for the 4th of July fish fry at Casa Ybel — chopping down a dozen cabbage palms to feed the 50 or so people. Later Francis became a civic leader, serving on the original Sanibel City Council.

John Bailey passed away a few years ago, but Francis and Sam still run Bailey's General Store, now more than 100 years old. What's its secret?

"I think of Bailey's not as the Bailey family store," Francis tells us, "but as the peoples' store, the island's store." Sam agrees. "My brother and I always say we are not good businessmen, but we believe in the island. So the island believes in us." Every month a chain tries to buy them out. At one time there were three agencies all trying to buy the store for Publix. Francis and Sam say they'll never sell.

Sam misses the freedom of the old days, especially the skinny-dipping and the fishing anywhere and everywhere.

Francis resents some of the newcomers, the ones that "come in, spend a day, and think they're experts on everything. But most of the people here want the same thing we old-timers have always wanted. And that is to protect and preserve the island."

Beaches

S troll along the shore. Slowly take in the changing beauty of the Gulf and sky, and the changeless beauty of the land. Or lie at the water's edge, and be soothed by the slow, rhythmic wash of the surf. Whatever you do, a few hours on the beaches of Sanibel and Captiva will refresh your mind and renew your spirit.

Our beaches are wild, as nature intended. Tropical birds and sandpipers stroll alongside you. Pelicans glide overhead; dolphins cruise offshore. Buildings are kept far away, tucked behind the palms and sea oats. The sand is never raked, so shells, seaweed and sea life wash up and back with the tide.

Out at sea the Gulf of Mexico can be turbulent. But the island waters stay calm, with rarely anything more than gentle waves and slowly curving breakers. Off-shore disturbances are evident only by a quick tumble of small waves. With a pair of binoculars you can often see an upheaval on the horizon, while only its minia-ture is at your feet. There's rarely an undertow; never a long, heavy surge.

The sea floor has a gradual slope — in many places you can wade out 500 feet before it's over your head, even at high tide. There are no holes or sudden dropoffs; the bottom is always clean white sand. And you can swim year-round. The Gulf water averages 71 degrees in winter; 84 in summer.

Island beaches aren't ruled by boom boxes and beer. Our beach music is the natural symphony of seagulls and surf. Our drug of choice is the purest intoxicant of all — total, complete relaxation. And, no, you don't have to hold your stomach in. No one does, and no one cares.

Gulf beaches

The southern and western shores of Sanibel and Captiva are actually one continuous 16-mile beach. But the curving shoreline creates six distinct areas. Each has its own public access point, with a parking lot and, in most cases, basic restrooms. Parking costs 75 cents per hour and is strictly enforced.

The narrow **Lighthouse Beach** *(at the east end of Periwinkle Way)* is the easiest beach to get to from the causeway, and the most crowded on the islands. During season, the park-ing lot fills by 10 a.m. Boardwalks run out to the beach, through arching mangrove roots that often form canopies overhead. You can clearly see Ft. Myers Beach from here, three miles across the bay. Look closely to see Bonita Beach and sometimes Naples farther down

Beach Science

Beaches have three distinct zones, each with their own purpose:

The first zone is along the water, the area where the waves break on the shore. This **intertidal surf zone** is a rich wildlife habitat. Snails, crabs, clams and sand fleas live here. Small fish swim up here to feed. Birds dine here too, darting in and out with the surf.

The **middle beach** is where most people throw down their beach towels and pitch their umbrellas. Too dry for sea life but too salty for plants, this area functions as a sand bank. Mother Nature deposits and withdraws sand from here — piling it up during one storm, carrying it back out to sea during another. Sea turtles nest on the middle beach, which keeps their eggs safe from the saltwater that can destroy them.

The **dunes** are the low ridges of sand at the top of the beach. They are the islands' security guards, protecting them from storm surges. The plants that grow here — beach morning glory, beach plum, seagrape, sea oats, cactus —have roots that grow into a tangled net, which keeps the sand from eroding when the sea rises to this height during storms. Even when the water comes over the dunes, the plants' stems and leaves dissipate the waves to minimize damage. A different environment from the rest of the beach, the dunes are home to lizards, snakes, tortoises and rabbits. But be careful walking here — the dunes are also home to millions of prickly, painful burrs.

Two **tides** rise and fall most every day, caused by the attractive gravitational force of the moon and, to a lesser degree, of the sun. As the earth rotates, the earth's deepest water (in the southwest Pacific Ocean) comes to be in line with the moon's pull. A small amount of water is pulled closer to the moon, and moves out around the globe. But tides in the Gulf of Mexico don't always follow this pattern. The Gulf has narrow openings on each side of Cuba that allow Atlantic waters, which are on a different tide schedule, to flow into it. The Gulf and Atlantic tides occasionally cancel each other out, resulting in no tide at all.

Red tide is produced by a reddish single-celled organism related to the Noctiluca. When these become abundant, outbreaks cause the death of fish and other creatures, which get cast up onto the shore to decay. (Don't touch these dead creatures; the organism can be harmful to humans as well.)

the coast. Be careful swimming; the strong, swift current can carry you out to sea. But this is a great shelling spot, the best place in the world to find the tiny wentletrap. While you're here walk up to the still-operating lighthouse and wander down the shady Gnarly Woods nature trail, a jungle of palms, seagrapes and other tropical foliage. The trail is hard sand, perfect for bikes and joggers. A separate parking lot serves the adjacent fishing pier.

Gulfside City Park *(also known as Algiers Beach, east of the Casa Ybel Resort, off Casa Ybel Rd. at the end of Algiers Dr.)* features a long, shaded picnic area. A portion of refuge land borders the dunes here, distancing the beach from the resorts that line the rest of this area. The Gulfside Park Preserve hiking trail is within walking distance, off the bike path off Algiers Drive.

There's plenty of room to stretch out at the **Tarpon Bay Road Beach** *(Tarpon Bay Rd. at West Gulf Dr.)*. This wide beach has good shelling and a surprising amount of wildlife. The parking lot, with special spots for recreational vehicles, is a short hike away on Tarpon Bay Road. Walking to the beach is safe from traffic; just use the bike path. The beach entrance has restrooms, a water fountain and a small outdoor shower.

Farther down West Gulf Drive, past Rabbit Road, are the seven **West Gulf Beach**

Top: The historic Sanibel lighthouse sits just a few feet off the beach
Below: A mother and daughter play cards at the Lighthouse Beach
Bottom: Sitting in the surf at Gulfside City Park

Boogie boarding on Bowman's Beach

Sand matters. As a rule, the closer the beach is to the lighthouse, the softer the sand. The very softest sand is at the Gulfside end of Buttonwood Dr. (below left); the powder is like that of a cigarette ashtray. The sand is coarser on Bowman's Beach (below middle). Captiva's beaches (below right), which are restored, are roughest of all; their sand has been dredged up from offshore.

Access Points. With no public parking, these remote beaches are accessible to visitors only by foot or bicycle.

One of the best beaches in Florida, **Bowman's Beach** *(five miles west of Tarpon Bay Rd., off Sanibel-Captiva Rd. at the end of Bowman's Beach Rd.)* has no hotels, homes or other development; it's just beach, beach and more beach. Two shaded parking lots have ample parking, again with special spots for RVs. A sand path takes you through Bowman's Beach Park, over a footbridge and out to the water. The beach runs for miles in both directions and is rarely crowded. The park here is the island's best spot for picnics and birthday parties, with dozens of picnic tables, scattered in three distinct spots, most with a barbecue grill (few people find the ones across the bridge, off to the left). Facilities include large restrooms, changing booths, a bike rack, two overhead showers and two foot showers.

A crowd gathers every night to watch the sunset at **Turner Beach** *(at the divide between Sanibel and Captiva, alongside Sanibel-Captiva Rd.).* The sky turns brilliant colors. Actually two distinct beaches, Turner Beach is divided by Blind Pass, the narrow cut that separates Sanibel and Captiva. (Often the pass is closed, but plans are to keep it open.) Each side has a parking lot and Port-A-Potty restrooms. Big and wide (sand continues to pile up here), Turner Beach is popular with fishermen and shellers. The gray, packed sand is filled with tiny shells. On the Sanibel

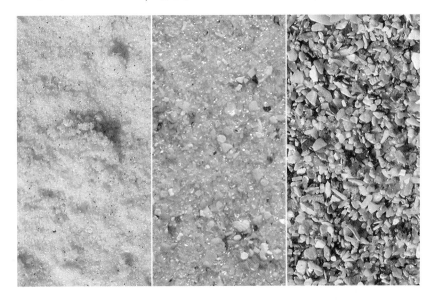

Building a Sand Castle

A baby can have the time of her life by just piling up a mound of sand. Two-year-olds will be thrilled to make a tall tower. But want a real sand castle? Follow these steps:

1. Be prepared. Bring these tools with you to the beach: Pails or buckets, small shovels or spades, plastic knives and spoons and a spray bottle.

2. Pick a good spot. Pick a level spot that is not too far from the water, but not so close that the incoming tide washes away your work before it's finished. Pour several buckets of water on your spot, then firmly pack a sand pile a few feet high, soaking it with several buckets of water. This will be the base for your castle.

3. Use wet sand. You can't make a sand castle with dry sand. You need lots of water to glue the sand together and let you sculpt and mold it into shapes.

4. Use towers. Once you've made the base, build your main tower — a tall, tapering tower with a wide base and narrow top. First, make a big pancake of very wet sand several inches thick. Gently add it to your tower and blend it in. Then make the next pancake, slightly smaller. Repeat the process until your tower is the height you want.

5. Build out. Use a combination of towers and walls. Cluster small towers around the central one in a pattern; make them anchor points for your walls. Make "dribble towers" with nearly liquid sand, dribbling it between your fingers. To build a wall, scoop up wet sand with both hands, squeeze out the water, and place it where you want your wall. Make a high wall by stacking one clump atop another, thick at the base and narrow as it rises.

Arches are simply walls with holes tunneled through them. After your wall is built, gently tunnel your way through at the base. Enlarge and shape the opening into an arch by shaving off thin layers of excess sand. Use plastic knives as shaping tools. Ramps are also modified walls — make them steep for staircases or gently sloping for walkways. Carve steps into a ramp's surface by using a straight-edged tool.

6. Focus on details. Decorate your creation with shells and other found objects (pen shells make good drawbridges). Keep your work from drying out by gently misting the castle walls and fine details with a spray bottle.

side, there's often a two-foot shelf of shells about 20 yards from the water. Swimming is dangerous on the Captiva side; the water gets deep quickly and there's a strong undertow.

Another great sunset beach, **Captiva Beach** *(at the end of Captiva Dr., Captiva)* is the northernmost public access point on Captiva, and borders the South Seas Resort. The public parking lot here is tiny (maybe a dozen spots), with no restrooms or facilities. Don't be intimidated by the South Seas Resort beach chairs you see here; the beach is public land, open to all. The best time here is the 4th of July, when South Seas launches fireworks over the Gulf.

Bay beaches

The **Sanibel Causeway** beaches *(on three man-made islands along both sides of the road)* are fine for swimming, fishing and picnicking. Windsurfers love it here. The water is shallow, so even young kids can splash around without fear. You can park your car anywhere you like, at the water's edge. Bring a chair to watch the windsurfers, boats and dolphins, or to watch the sun set over Sanibel.

The bay side of Sanibel is generally covered by mangroves, but there are some sandy areas. The north end of **Buttonwood Drive** (near the lighthouse) has a skinny

Top: A pick-up volleyball game at the 'Tween Waters Inn. **Above:** Surfing off Captiva's Turner Beach. **Below:** Taking it easy on the Sanibel Causeway. **Facing page:** Afternoon clouds on Bowman's Beach.

Nude No More

Sanibel native Sam Bailey, 78, says he and his brothers used to go skinny-dipping at the beach all the time. "We'd strip off our clothes as we walked, and toss them into the grass when we got to the dunes." Growing up at the family homestead on Periwinkle Way, the beach was just a short walk out Donax Street.

"There wasn't anything brazen about it," says his brother Francis, 80. "Sometimes we'd just be walking along the beach and decide to shuck our clothes and cool off."

Other islanders carried on this tradition. Talk to a Sanibel resident long enough, especially over a couple of beers, and eventually you'll hear a tale about skinny-dipping back in his or her younger days. The popular spot in the 1970s and 1980s was Bowman's Beach, especially at the remote western end at Silver Key. Sure, it was against the law. But it was discreet, innocent, and harmless. Island police looked the other way.

But now those carefree times are gone. In the 1990s, the open-minded attitude on Sanibel began to get national publicity. Web sites touted Bowman's Beach as an "official" nude beach. Some said it was a gay beach. Visitors began coming to the Silver Key area specifically because of its nude or gay reputation. Locals and visiting families began to stay away. Stories of lewd exhibitionism and sexual aggression began to appear in island papers.

Then tragedy struck. Late one evening in 1998 an elderly woman went for a walk in the woods at Bowman's and was attacked by a nude man. The incident outraged island residents, and though the attacker was never directly associated with the nudist crowd, the free-spirited days of old were clearly over. The Sanibel police began routine ATV patrols of all beach areas. They also began enforcing the anti-nudity regulations — laws that never really seemed right for the islands before. The awful stories went away, and Bowman's again became known for its natural — instead of *au naturel* — attractions.

Today there are no areas where public nudity is allowed on the islands, officially or unofficially. Don't expect officers to look the other way anymore; they are under orders to arrest nude sunbathers. And don't believe rumors that the laws can't be enforced. Though cases in Florida have been dismissed against nudists arrested under state statue 800.03 (Exposure of Sexual Organs), Sanibel Police arrest offenders under a different statue, 877.03 (Breach of Peace and Disorderly Conduct). And the arrests stick.

stretch of sand (barely walkable); ditto for the end of **Dixie Beach Road.** Your best bet: the end of **Bailey Road,** next to the causeway. This shady spot is popular with boaters. Captiva's only bay beaches are small, man-made areas at the 'Tween Waters Inn and the South Seas Resort.

Beach tips

What to bring to the beach

Figuring out what to take to the beach may seem like a no-brainer: a swimsuit and a towel. But don't forget sunscreen, a cap (or hat) and polarized sunglasses. Other ideas: money for parking; a cooler with drinks and snacks; a book or magazine; beach shoes; your camera; a snorkel and mask; sand toys for kids (pails and shovels should be at the top of your list); a plastic bag or net to collect shells; a plastic shell and/or bird I.D. guide; a beach towel or quilt to spread out on the sand; a beach umbrella; and beach chairs. What not to bring: a radio or boom box. No one uses them here; they clash with the back-to-nature ambience.

Many stores sell beach supplies; many resorts provide equipment for guests. You can rent beach goods, too. Rental companies include Island Rental Service *(472-9789),* Sanibel Rental Service *(2246 Periwinkle Way, Sanibel; 472-5777)* and Jim's Rentals *(11534 Andy Rosse Ln., Captiva; 472-1296).*

Sea foam is common on Bowman's Beach whenever Clam Bayou, an estuary a mile west, is open to the Gulf. It's created when debris from living plants washes into the sea and dissolves. The organic material alters the composition of the water just enough to allow bubbles to form during a strong wind.

Kayaking at Casa Ybel. Gulf-front resorts rent a variety of boats and other beach equipment.

Sunset times
(p.m., first to end of month)

January	5:46–6:09
February	6:10–6:28
March	6:29–6:44
April	7:45–7:59
May	8:00–8:16
June	8:17–8:25
July	8:25–8:15
August	8:15–7:48
September	7:47–7:15
October	7:14–5:45
November	5:45–5:34
December	5:34–5:46

Sea turtles nest on island beaches in the summer

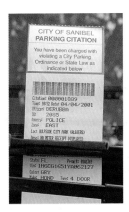

Beach parking is 75 cents an hour, and is enforced with a passion. Parking tickets are common.

Other tips

1. Tide times change daily. Tide charts (available at island marinas) show tides for the Sanibel lighthouse. Add time as you move farther west or north — an hour for tide times for Bowman's Beach and Turner Beach, up to 90 minutes for beaches on Captiva.

2. Alcohol is allowed on Sanibel beaches during the day (one hour before sunrise to one hour after sunset) from Dec. 15 to May 15, and at all hours during the summer. Alcohol is prohibited on Captiva year-round.

3. Pets are allowed on Sanibel beaches if leashed. No pets are allowed on the causeway or Captiva beaches.

4. Watch the water. You may see a dolphin arc out of the waves, or the silvery flash of passing tarpon.

5. No island beach has a lifeguard; you swim at your own risk.

6. Sculpt the sand to your body before you lay out your towel. In other words, make yourself a sand pillow, and dig little depressions for your elbows, heels and bottom. It's incredibly comfortable!

7. Keep snacks out of sight. When you're not looking, birds will invade your chip bags and cookie boxes.

8. Spit on the inside of your goggles and masks (and rub) to keep them cloud-free.

9. Keep an eye on your kids. Young children often get lost on the island beaches, and sometimes become dehydrated or seriously sunburned. They wander far away from their parents, then forget which way they came from (easy to do, since the natural wilderness has no landmarks.) If your child gets lost, call the Sanibel Police at 472-3111. On Captiva call the Lee Co. Sheriff at 477-1200.

10. If you drive to the beach, leave a few bottles of tap water in the car. When it is time to leave, you can pour the water over your sandy feet before you get in the car.

11. Put your valuables in the trunk before you get to the parking lot. Serious crime is rare on the islands, but thefts from beach parking lots are common. Thieves wait in vans with darkened windows, watching until they see people put purses or other valuables into their rental car trunk. Once the victims go innocently off to the beach, the thieves break into the trunk.

12. Islanders call Gulfside City Park "Algiers Beach." Algiers was the name of an old ferryboat that was pulled ashore here in the 1950s and converted into a home. Sanibel officials tore down the abandoned boat in 1980, and converted the back yard into Gulfside City Park. Rumor has it there are still pieces of the boat laying in the woods.

MERISTAR

Spectacular Sunsets

Each evening people gather at the island beaches, cameras ready, awaiting the nightly show. But photographs don't do it justice; an island sunset can be breathtaking. Just after the sun disappears the clouds turn a bright silver. When the sun's on its last gasp — when it's just a tiny sliver of bright gold — the light flattens out and spreads across the horizon. Keep your eyes peeled for the elusive green flash right at the horizon as the sun dips below the edge. About a half hour after the sun disappears, turn around and watch the eastern sky for something just as interesting — a dark purple haze rising from the east. This is the earth's shadow. When the shadow circles around to cover the entire sky, it becomes dark.

The sky changes colors as the sun goes down. The sun's light passes through more and more air, which robs the white rays of their blue color. The sky turns orange and red at dusk because the dust in our air reflects these colors. Sometimes the clouds above the islands are ideally situated to reflect the rays straight down. The resulting "pink light" throws a unique rosy cast over the ground. But it happens rarely; only a few times a year.

Of course, there really is no such thing as the sun "setting." During a sunset, the sun is for the most part standing still, while the earth rotates — the lands west of us rise up to block our view of the sun. What we are really watching is an "earth-rise."

30 Beach Activities

Simple pleasures make the best memories, especially at the beach: just lying on the sand can be a total escape from the daily routine. But besides the obvious — shelling, swimming, sunbathing — what else can you do on an island beach? Here are 30 ideas:

1. **Build a sand castle.**
2. **Read.**
3. **Play cards.**
4. **Take pictures of each other:** parents of kids, kids of parents or grandparents (shown at right), dates and spouses of each other (for fun, pose beefcake and cheesecake style). For the best shots go in the early morning and have your subjects face the sun.
5. **Play on floats.**
6. **Hunt for sea life.**
7. **Fish.**
8. **Surf on a boogie board** (or on a surfboard, if the waves are big enough).
9. **Ride a skim board** along the surf line.
10. **Fly a kite.**
11. **Sleep.**
12. **Have a picnic.**
13. **Jog.**
14. **Sit in beach chairs at the water's edge,** letting the waves wash over your legs.
15. **Take a long walk** and look for dolphins.
16. **Put on a mask** (and snorkel, if you like) and explore right at the edge of the surf, where the waves hit.
17. **Make a giant magnifying lens** by suspending a puddle of water on a piece of plastic wrap. You can use it to explore all the wonders of the beach, from the multicolored grains of sand to the tiny insects floating on the surface of the water.

18. Sculpt a face in the sand (such as shown at right), or a whole person, or mermaid. Decorate it with shells and beach debris. Use seaweed for hair, moon snails for eyes, pen shells for a crown.

19. Scan the sea floor. Cut off the tapered portion and bottom of a large plastic soda bottle so that you are left with just the middle section. Stretch a piece of plastic wrap over one end and secure it with an elastic band. When you push the bottle through the surface of the water, you will have a clear view of any objects sitting on the sand below.

20. Join a naturalist from Tarpon Bay Recreation Center *(900 Tarpon Bay Rd., Sanibel; 472-8900)* on a walk along the shore of the Perry Tract of the J.N. "Ding" Darling National Wildlife Refuge. Walks are at 8:30 a.m. Tuesday and Friday. The cost is $5 per person (children 5 and under free).

21. Go to the beach right at dawn. The ultimate relaxing, natural experience.

22. Go at night. "I love walking on the beach in the middle of the night," says Brianne Meyers, 21, who grew up here. "There's no one else around, and it's very quiet and peaceful." Take a flashlight and you may see a great deal of activity on the sand and in the water. Many crab species roam the beaches. Point the light steadily at the water right after dusk and you may attract many swimming creatures, especially worms. Turn off the light and splash the water and you may see the star phosphorescence, pin-points of light caused by single-celled organisms called Noctiluca. If you're here in the summer or early fall, watch for loggerhead turtles to climb onto the beach to lay eggs.

Or play a game. Bring a **(23)** frisbee, **(24)** football, **(25)** a baseball and a pair of gloves, or **(26)** one of those zany plastic things. Organize **(27)** a pick-up game of volleyball (resorts often set up nets). Or try one of these simple beach games:

28. Balloon toss. This game puts a smile on everyone's face. *Equipment:* Full water balloons (fill them up at a beach restroom, or bring them to the beach in a Hefty bag). *Number of people needed:* Any even number. *Description:* Players divide into two-person teams. Teammates stand five steps apart, facing each other. Players toss a water balloon back and forth. After a player catches the balloon, he or she takes one step back, increasing the distance to the teammate. Players continue to move back until one player fails to catch the balloon (i.e., when it breaks). That team is out. The last team left is the winner.

29. Towel toss. *Equipment:* One beach towel and beach ball (or bean bag or other ball) for every two people. *Number of people needed:* Any even number of at least 4. *Description:* Players divide into teams of two. Teammates face each other, holding their towel between them. The ball is placed in the center of the towel. Each team pumps their towel to toss their ball in the air, catching it as it returns. The team that does the most tosses without missing the ball, or whose ball goes the highest, is the winner.

30. Name that cloud. *Equipment:* Blanket (optional). *Number of people needed:* Any number. *Description:* Spread the blanket on the sand, then lay on your back and see what shapes you can make out of the clouds. *Note: Play this game with children. You'll be surprised how imaginative they can be.*

Beach girls. Nicole Ogden and Julia Leal, both 7, play tic-tac-toe

An Afternoon at Bowman's Beach

By Julie Neal

Just getting to Bowman's Beach is fun. As I walk across the footbridge that crosses Old Blind Pass, I pause and look over. On my left there's a wood stork patrolling the shore, trolling for fish. I look down and see an alligator, silently sneaking by underneath me.

I head to the beach along a sandy path. The upper dunes are a beautiful dry desert. Seagrape trees, prickly-pear cactus and black-eyed Susans grow with abandon, all close to the ground, braced for the next storm. Many lizards, and one snake, scurry in the sand.

Now creeping grasses and vines appear, a welcome mat stretching out toward the water. A few sea oats wave in the breeze. I'm still a hundred yards from the water, but many shells lay on the sand. Perhaps they've been here since 1960, the last hurricane.

Finally I'm to the beach. Crossing over the last dune and seeing the shoreline spread out before me, miles in each direction, always stops me in my tracks.

Bowman's is big. It goes as far as you can see, left and right. The sand is soft; white and gray mixed together. Shells are everywhere.

Nearly a hundred birds are gathered at the water's edge — laughing and ring-billed gulls, terns and plovers. They preen and stand around. Then they look at each other and, as a group, lift up and fly over to a sandbar to feed. Two willets have a screeching fight in front of me. They peck at each other, and, like dancers, rise and swoop at each other's backs. A few feet away tiny sandpipers race the breaking waves, darting down into the wet sand to probe for food between each wave's crash onshore.

There are only a few dozen people here, spread out in both directions. Couples, young and old, are holding hands, strolling by the water's edge. A young woman is putting tiny shells in a pill bottle. She's the only person doing the Sanibel Stoop. A father and son are fishing, using shrimp as bait. When they bait up, 20 or 30 gulls fly up and hover around them, like a scene from "The Birds." The little boy waves to the birds and laughs.

A woman bicycles by, but is having trouble in the soft sand.

Two teenage boys are chest-deep in the water, tossing a football back and forth. Their beefy, crew-cut dad joins in. The mom, blonde and thin, floats nearby on a red raft. Now the boys and their dad are wrestling in the water, laughing. The mom is laughing, too, splashing them with her legs. They try to tip her off her raft. But she escapes, barely.

I notice I'm not the only one watching people. So are the gulls. When an old man gets up off his beach towel, one secretly sneaks up and pokes around the towel for food. Later, a woman leaves her blanket behind, and another gull helps himself to her bag of chips.

Small fishing boats pass by every few minutes. Each has only two or three people on board. None seems to be in a hurry.

A mom and her little girl walk by with a happy, crusty golden retriever. The dog's fur, even its face, is covered with dried, chunky sand.

The perfect playground

Before I even make it to the water, my 7-year-old, Micaela, runs ahead and makes friends with another young girl (a complete stranger — imagine adults doing that). Dressed in a bathing suit and big yellow rubber rain boots (her idea), Micaela starts building a sand castle with her new friend at the water's edge.

Then a dolphin surfaces, not more than 15 feet away. Micaela and her friend get up and chase the dolphin down the beach. Over and over again, the dolphin breaks the surface, disappears, then pops up a minute later, farther down the beach.

"Blowhole!" the kids scream, and run down the beach after him.

"Blowhole!" Run. "Blowhole!" Run. "Blowhole!" Run.

This goes on for nearly 10 minutes, until the kids are just little specks in the distance, far up the beach.

Later Micaela finds a coconut floating in the surf. She tries to open it up with the edge of a pen shell, but loses interest quickly. She kicks off her boots, gets in her ring, grabs her alligator puppet and runs into the water. She carefully positions herself so the waves break over her. Sunscreen — "waterproof" sunscreen — is washing off with every crash.

Next she brings me a huge dead horseshoe crab, big as a dinner plate. Her eyes arc wide. "The beach is covered with shells that are alive!" she tells me. And she's right. I walk with her and we toss dozens of live blue-eyed scallops, as well as sea urchins and sea anemones, back into the surf.

Micaela walks up on a flock of willets. They squawk and bob their heads. She squawks and bobs her head.

She sees a fisherman and goes up to say Hi. He gives her an 8-inch-long silver fish; she puts it in her big red bucket. She names the fish Bob.

Now she's digging holes in the sand, filling them with water from her other bucket. She sings softly as she plays. Sometimes she sings to Bob. Sometimes she has a plastic dinosaur talk to Bob. She picks up Bob and looks at him, breaking every "catch and release" rule known to man.

Still digging away, Micaela uncovers hundreds of coquinas, each a different color. She picks them up by the handful and lets them burrow down to her fingers.

I tell her it's time to leave — it's nearly dark. But wait! "There's a big hole!" Micaela runs off to a 3-foot-deep hole someone else has dug. She climbs in, and only her head is visible. She waves to me. Now we're ready. But wait! A crab hole! Micaela tries to dig the crab out, being quiet "so the crab won't leave."

Finally, she's ready to go home. She tells Bob she's letting him go, and tells me he looks happy to hear it. Then she pours him in the surf, and watches him slip away.

Shelling

*W*alking along the beach, looking for beautiful, free treasures, your mind escapes from the pressures and headaches of the world back home. You are moving with the shore birds, walking along with them as they skitter along the edge of the waves, poking the wet sand for food. You occasionally look up and scan the water for signs of a dolphin.

As Robert Louis Stevenson once said, "It is perhaps a more fortunate destiny to have a taste for collecting shells than to be born a millionaire."

The only problem is you can't stop. Shells are everywhere on the beaches, each one waiting patiently for you to find it. You just know a gorgeous shell — maybe a junonia — is right up there in the next stretch of sand. Anything is possible, so why stop right this minute? "Yeah I'm coming… just a second…"

A shell is a lovely gift from the deep, with a graceful design and intricate detailing equal to the prettiest flower or butterfly. Every single one is different. Most are conveniently sized to fit in your pocket, and, later, that vase in your living room.

Newcomers want to pick up nearly every shell they see. Old-time shellers stop for fewer shells, only the unbroken, unblemished ones.

Shelling is good on Captiva, and the out islands have great shells because few people visit them. But the Mecca — the ground-zero of the shelling universe in the Western Hemisphere — is Sanibel. The only spots better, in all the world, are the Sulu Islands of the Philippines and Jeffrey's Bay on the southeast tip of South Africa.

Sanibel lies east to west, sticking out into the Gulf about 10 miles. Shells from the Caribbean Sea that would normally roll on by instead roll right up on its beaches. The pattern of water circulation in the Gulf combines with the gentle slope of the sea floor — and, in the winter, the prevailing northwest winds — to bring up thousands of great specimens every day.

Conchology 101

Shells are more than just pretty baubles. They are the exterior skeletons — exoskeletons — of a group of soft-bodied animals called mollusks ("mollusk" means "soft-bodied"). An exterior skeleton is vital to these creatures. It gives them shape and rigidity, and provides them with protection, and sometimes camouflage, from predators.

Facing page: Piles of shells wash up on the islands with every high tide

Facing page: Shelling at the Lighthouse Beach

Have a treasure hunt with your family. Make a list of 10 shells to find, then compare treasures back at your beach umbrella.

Live shelling is illegal on Sanibel. The ban extends ½ mile offshore and includes all refuge waters. Sand dollars, starfish and sea urchins are included. The fine for a first offense can be $500 and 60 days in jail. On Captiva you can take two live shells per species per day.

Stop and search one spot for a while (do the "Sanibel Stoop," as demonstrated by islander Cindy Sitton, below). Often a shell will be too small to see with only one glance.

There are between 50,000 and 200,000 species of mollusks. Estimates vary depending on who's guessing the number of undiscovered species.

Types of shells

Four types of mollusks are common on Sanibel and Captiva. Hinged shells, such as scallops and clams, are called bivalves — they have two sides, or "valves." The animal is attached inside. Spiral shells, such as conchs and whelks, are univalves — they have only one "side." They can move within their shell freely. A third group, gastropods, are single-shelled snails that are not spiral shaped, such as a slipper shell.

The fourth group probably doesn't come to mind when you think of mollusks, since they have no shell: octopus and squid *(Cephalopoda sp.).*

Bivalves protect themselves with strong muscles that keep their hinges shut (just try to open an oyster bare-handed.) Most univalves and gastropods stay safe with an operculum, a trap door which closes across the shell's opening.

There are more bivalves than anything else. Univalves often eat bivalves. The spiral shell will grab its hinged relative with its foot, drill a hole into the bivalve's shell with its file-like teeth, and pull out the meat. Some univalves do it differently, injecting a muscle-relaxing chemical into the bivalve, then prying it open and munching away.

Mollusks not normally found on the islands include tusk shells *(Scaphopoda sp.),* which have a single shell, but no coil; chitons *(Polyplacophora sp.),* which have a row of eight overlapping plates; neopilina *(Monoplacophora sp.),* with single shells that fit over their bodies like cups; and the deep-sea worm-like aplacophora *(Epimenia australis).*

Growth and life

When a mollusk hatches from its egg it already has a tiny shell. As it grows, it expands its shell by absorbing material from the water, either directly or through the food it eats. Then the mollusk creates, and secretes, calcium carbonate, which hardens around it.

You can learn a lot about a mollusk's life by looking at its shell. Many have scars — healed-over breaks and chips — from battles with predators. Others show a color change, because of variations in diet or water chemicals. The shells of older mollusks are often thick, or dull in color.

Shell shapes

Each species makes a shell that is, in most cases, unique to it alone. Shell shapes have evolved to make each animal better suited to its habitat. A mollusk that burrows needs a shell that will move through wet sand easily — a smooth, slender, tapering shell, narrow at the front, with no impeding projections. A mollusk that needs lots of camouflage may have evolved with a spiny or irregular shell, letting it catch and hang onto all sorts of encrusting small creatures and plant life.

Each species is destined genetically to develop the same type of shell its ancestors did. But, just as with humans, there are many distinct differences. Food, climate, environment and the mollusk's particular heredity all play their parts in making each shell unique.

Sanibel Brownie Nicole Horton, 7, examines her shells with Girl Scout leader Lisa Williams

Color and shine

The color of a shell is determined by genes, or at times, by the diet of the animal. The color often camouflages the shell in the water. Some pigments, such as the yellows and reds of beta carotene, help strengthen the shell.

Some shells are polished by the animal living inside. A portion of the mollusk's body slides up around the shell, like a cape, to help the animal crawl or feed. Rubbing its shell as it moves, the mollusk smoothes and shines the surface, keeping the shell free of growth and encrustations, and from dulling by chemicals in the water.

Record setters

The biggest American marine mollusk, the Florida horse conch *(Pleuroploca gigantea),* can be up to 2 feet long. The world's largest marine mollusk is the giant clam *(Tridacna gigas)* of the southwest Pacific; the biggest ever found measures 4 feet, 7 inches. Second place belongs to the Australian trumpet *(Syrinx aruanus),* at 2 feet, 6 inches.

Tips for good shelling

It's not rocket science to find shells on Sanibel or Captiva. They're everywhere, on every beach; you'll find many great specimens with just a five-minute walk. But you'll double or triple your bounty if you follow a few simple tips:

1. Go at low tide, after a storm. You'll find the most shells at low tide, which usually happens twice a day.

Should I stay or should I go? An oyster doesn't travel in its adult life. It attaches itself to a rock or other shell when young, and lives the rest of its life there. Scallops, however, migrate in large numbers to find richer feeding grounds. A scallop moves by squirting jet streams of water out of its shell.

Facing page: Get the scoop. Talk with other shellers on the beach. Ask what they have found that day and where the good spots are. Shellers love to talk, and will usually share secrets about their favorite spots, methods and contraptions.

More shelling tips

■ Look for round shells, such as nutmegs, cones and olives, right at the surf break — the ledge in the sand, a few inches to a foot high, where the waves break on shore.

■ Bring a large baggie to the beach to store your sand dollars. They'll break apart if you mix them in with your shells.

■ When you bring your shells back to your room, rinse them thoroughly in the sink. They may look clean, but inside there's often sandy muck that will start to reek in a day or two if not washed out.

■ Buy a plastic-coated shell guide to take with you to the beach to identify the shells you find. Pick up one at the Shell Museum gift shop, the refuge visitors center bookstore or one of the island book shops.

■ Put on a diving mask and snorkel in the water, just a foot or so offshore. You'll find more shells than just looking in the water while standing up.

■ For variety, go out once and only look for tiny shells. Bring a pill bottle to put them in.

Facing page: Angel wings are a top find, especially if you discover two matching halves

Rarer varieties of shells are more common then. The receding water gives you a lot more beach to search, and it exposes sand bars covered with scores of shells, including many that never make it to the beach. The lowest of low tides is during a full moon. Right after a storm is another great time. Big waves loosen a lot of shells from the sea bottom and wash them up. After a big tropical storm, usually in the summer, the waves will bring up rare and unusual shells for days.

The best time of all, then, is low tide after a storm, ideally during a full moon.

2. Bring a container. You'll find many more shells than will fit in just your hand, so bring a plastic bag — the ones from a grocery store work great — or a bucket. Another good choice is a milk jug. Cut off the top, but keep the handle, and you'll have a bucket with a grip. Small net bags are great, but they may start to smell after awhile. Forget about paper bags. Wet shells will break through the bottom.

3. Slow down. Move at a snail's pace (sorry, we couldn't resist). Stand in one spot for a while just to look more closely. Sometimes a wave coming in will uncover a shell at your feet. Or a shell may be so small that you will spot it only after a second glance.

4. Wear a hat. Going shelling is an easy way to burn your head. Wear a cap or hat. (Devoted shellers even wear hats at night — *miner's* hats, complete with lights.)

5. Wear shoes. Wear shoes when you walk in the surf — sneakers, beach shoes, any shoes that can get wet. Sections of the beach may be literally covered with shells, making it hard to walk with bare feet. Shards from broken shells can really hurt, and pen shells, which are common here, have razor-sharp edges that can cut your feet.

6. Respect live shells. Some mollusks move so slowly you'll think they're dead. Many hide in their shell and "play possum" when you pick them up. But take only unoccupied shells. Snails often curl up into their shells and close their operculum that keeps them safe inside. If you see that door or a fleshy foot sticking out, the snail is still alive. If you have any doubt, assume it's alive and place it gently back in the sea. It's against the law to take live shells, and police are out there, both on the beach and in boats.

What you'll find

Angel wing

Angel wings (*Cyrtopleura costata*) are bright white. Some have a touch of pink on their interior. A boring clam,

When you bring your shells home, separate them by species, and put each into a separate bottle. You'll end up with a lettered-olive bottle, a fighting-conch bottle... a whole collection for your shelves and bookcases.

Many mollusks breathe through a snorkel, or siphon, when they bury themselves in the mud or sand

Coquinas are popular with children. Dig down a few inches, scoop up a big handful of sand, and you may find dozens of these little guys frantically burrowing down toward your skin.

the angel wing lives on the sea floor, digging into soft, thick mud by rocking back and forth. It is never found alive on the beach, and rarely washes up intact — a delicate bracket that holds the two halves together usually breaks apart in the surf. It is a top find among collectors.

Clam

Tiny **coquinas** ("ko-KEE-nahs") *(Donax variabilis)* live just beneath the beach surface at the surf line. Exposed by a breaking wave, they wiggle themselves back in the sand with amazing speed — it's "the Dance of the Coquinas." A favorite snack of gulls and wading birds, coquinas migrate up and down the beach to stay wet as the tide rises and falls. Usually striped, the shells can be purple, red, white or yellow outside; often a completely different color inside. They are common on the islands from July through October. Live coquinas have two small siphon tubes.

Its pattern reminds some people of a turkey, others of a zebra. What most folks call a **turkey wing** *(Arca zebra)* is easy to identify it by its brown-and-white, turkey-like coloring and its long hinge. The inside of the shell has scars that look like teeth marks. Like pen shells, jingle shells and jewel boxes, turkey wings sew themselves onto rocks offshore with a byssus ("BISS-us"), an anchor chain with thin, flexible strands. Those threads break during a storm, when the shells wash up onto the beach.

The **calico clam** *(Macrocallista maculata)* has a blurry, brown-and-white pattern. The **cross-barred venus clam** *(Chione cancellata),* one of the most common shells on the islands, is easy to distinguish by its cross-hatched ridges. Those who grew up before the age of CDs can identify a **dosinia** *(Dosinia discus)* by its thin, circular ridges that look like the grooves of a phonograph record. The similar **duck clam** *(Raeta plicatella)* is more of a triangle, with the grooves farther apart. The **ponderous ark** *(Noetia ponderosa)* has a black skin-like covering, called a periostracum, that often washes off as it washes up to the beach, leaving only the pure white shell. The shell of a **quahog** (pronounced "COH-hog") *(Mercenaria campechiensis)* makes a great car-key holder for your kitchen. Shore birds love to snack on the **sunray venus clam** *(Macrocallista nimbosa).* Finally, the **surf clam** *(Spisula solidissima)* is easy to spot by its tan periostracum.

Turkey wing

Calico clam

Cross-barred venus clam

Dosinia

Duck clam

Ponderous ark

Quahog

Sunray venus clam

Surf clam

Crown conch

Florida fighting conch

Florida horse conch

Alphabet cone

Conch

The **crown conch** ("kawnk," not "kawnch") *(Melongena corona)* has a row of sharp exterior spines, sometimes more than one, which gives it a crown-like appearance. Rarely found on beaches, it lives in mud flats, and often clings to mangrove roots, eating oysters. At low tide, the crown conch sits on the exposed sand, camouflaged by a covering of mud and slime. It holds a supply of water in its shell that allows it to "breathe" when out of the water. Look for many just off the shore in Clam Bayou or at the Red Mangrove Overlook in the refuge. This handsome shell is also called the king's crown conch.

The **Florida fighting conch** *(Strombus alatus)* is an aggressive, carnivorous mollusk. Pick up a live one and it will beat your hand with its foot, sometimes drawing blood. Set it down upside down, and it will turn itself over. Notice its bright blue eyes, located on half-inch stalks at one end of the shell. Live shells come near the beach during breeding season, but spend most of their time hundreds of yards offshore.

The official state shell of Florida, a **Florida horse conch** *(Pleuroploca gigantea)* can get as big as a horse. Well, it seems that way. One of the largest shells in the world, a Florida horse conch can be more than 2 feet long. But large ones are rare. Half-inch-long shells from young conchs are common on Captiva. The shell is orange outside — bright orange or yellow when young — and red or brown inside.

Cone

An **alphabet cone** *(Conus spurius atlanticus)* has a pattern of orderly, dark scribbles which, in an abstract sense, looks like the Chinese alphabet. A live cone stabs its prey with poison-tipped harpoons, each the size of the tip of a ballpoint pen, that grow from its teeth. The poison in American cones is not dangerous to man. The smaller **Florida cone** *(Conus floridanus),* also found here, has a more pointed top, and often a light-colored stripe around the middle.

Fig

The whitish-tan **common fig** *(Ficus communis)* is 3 to 4 inches long, and shaped like a fig. Similar in shape to a pear whelk *(Busycom spiratum)*, also found on island beaches, the fig has a fine crisscross pattern that sets it apart. Also called a paper fig, the shell is fragile. Few make it to the beaches intact, and those that do are easy to accidentally crush in your hands or shell bucket. A fig does not have an operculum.

Common fig

Jewel box

The **Florida spiny jewel box** *(Arcinella cornuta)* is white and has long spines, often worn down by the sea. If you find an intact shell, notice how the two halves combine to form a container shape. Live jewel boxes attach themselves to rocks or other shells with a byssus. Other shells with a byssus include turkey wings, jingle shells, and pen shells.

Florida spiny jewel box

Jingle shell

Jingle shells *(Anomia simplex)* are thin, soft, tiny and flaky. They come in a variety of colors, and are often used in shell craft. A scar inside the shell, where the mollusk muscle was attached, is said to look like a human baby's foot, sometimes even with a heel and toes. A handful of these delicate shells will jingle if you shake them.

Jingle shells

Junonia

Find a **junonia** *(Scaphella junonia)* and you'll get your picture in one of the island newspapers — it's that special. (Rumors still circulate that the Chamber of Commerce sometimes plants them on the beach.) This cream-colored shell has irregular brown spots that wrap around it. Most are about 3 inches long. Loved by locals because of its rarity and beauty, the junonia is the pride of Sanibel. You'll see images of it decorating signs, stores and restaurant menus throughout the islands. The live shell lives in deep water, miles offshore. It usually takes a good storm to wash one up onto the beach.

Junonia (also on overleaf)

Atlantic kitten's paw

Baby's ear

Atlantic moon snail

Apple murex

Kitten's paw

The **Atlantic kitten's paw** *(Plicatula gibbosa),* also called a cat's paw, has ribs which make it look like the paw of a small cat. It's common here, though usually you'll find only the top half. The bottom glues itself onto other shells and rock, and stays put even after the mollusk dies. Though it looks like one, a kitten's paw is not a scallop, and not related to the similar, but larger, lion's paw.

Moons

The beautiful **baby's ear** *(Sinum perspectivum)* has an opening so wide you can see the entire inner shell, which is said to look like the ear of a small child. The white outer shell may have yellow or tan accents.

Although harmless as a dead shell, in life the gorgeous **Atlantic moon snail** *(Polinices duplicatus)* is an acid-spitting predator — secreting a corrosive acid onto a clam shell to soften it up before boring into it. It can eat several clams a day. The moon snail shell has vivid colors, including blue spots that look like eyes — in fact, it's also called a shark's eye. Moon snails have been painted by many artists, including Georgia O'Keefe. Anne Morrow Lindbergh used to write poems about them when she stayed on Captiva. (A third type of moon is the **natica** *(Natica canrena),* a butterscotch shell with lines and spots, rarely found here.)

Murex

The **apple murex** *(Phyllonotus pomum)* is the most common murex on the islands. Usually 1 to 3 inches long, this sturdy univalve is round and bumpy, without sharp edges, so it washes up on the beaches in fine shape. It's usually light brown with a gray tint. A good spot to find a live apple murex — for viewing purposes only, remember — is on the back side of a sandbar, the side closest to the beach. It likes to dig in muddy sand. The operculum is round, and located in the middle of the shell.

The **lace murex** *(Chicoreus florifer*

dilectus) is a terrific souvenir, if you can find one in good shape. About 2 inches long, it has delicate spines and other frills. Most are brown; some are white or black. In the water, the live shell anesthetizes its victim (usually a clam) with a smelly yellow poison, which turns to purple in the sunlight. Because of its exclusivity, ancient royalty, including Antony and Cleopatra, used this liquid to dye their robes, giving us the color "royal purple."

Lace murex

Nutmeg

Common nutmegs *(Cancellaria reticulata)* look like nutmeg seeds. Usually brown, sometimes white, they have a beaded surface. Many nutmegs have scars. Often you can see break lines where the shell has been broken from crab attacks. They live in water 20 to 30 feet deep, so it takes a good storm to wash one up to the island beaches. But you may find one at the surf line, or perhaps out on a sandbar.

Common nutmeg

Olive

Lettered olives *(Oliva sayana)* are easy to find. This rounded oval shell is covered with intricate markings that look like crudely drawn letters. It has a point at the top and a few ridges at the bottom. Most lettered olives are very shiny, and are tossed onto the beach unbroken. An alternative form is the rare, yellow golden olive.

Lettered olive

Pen shells

A large, dark, brittle, spiny, unattractive shell, most folks ignore the **saw-toothed pen shell** *(Atrina serrata)*. But when you find one, pick it up and look at the outside. Live slipper shells, oysters, barnacles or other creatures may be attached to it. Break it open, and you might discover live baby murex and banded tulips. And notice the inside lining of the shell — it's iridescent.

Even uglier on the outside, with raised ridges, the **stiff pen shell** *(Atrina rigida)* hides an even more beautiful iridescent lin-

Saw-toothed pen shell

Stiff pen shell

Calico scallop

Zigzag scallop

Lion's paw

ing, in a rainbow of colors.

A pen shell mollusk creates its lining by creating layer after layer of thin nacre — the same material an oyster uses to make a pearl — as it builds its shell. Many pen shells wash up after storms, as they live just offshore in shallow water. Pen shells have strong hinges, so you'll often find both sides of the shell still connected.

Scallop

You'll find many different color combinations of the **calico scallop** *(Argopecten gibbus)*. You've seen its distinctive shape thousands of times — it's the symbol of the Shell Oil Co. A live scallop can swim freely, and can dart out of the way of a starfish, its main predator. It moves by opening its shell and clamping it shut, over and over again, squirting out jets of water to propel it along. Rows of blue eyes can sense light. Scallops served in restaurants are usually calicoes.

The top half of a **zigzag scallop** *(Pecten ziczac)* makes a good centerpiece for a necklace. The only part you're likely to find on the beach, the top is completely flat. It looks like a fan, with small ribs. It usually has a zigzag decorative pattern, and comes in a variety of colors. The rarely found bottom half is deep and concave, like a cup. (The similar Ravenel's scallop *(Pecten raveneli)* has more prominent ribs, and no zigzag pattern.)

The largest American scallop is the stunning **lion's paw** *(Lyropecten nodosus)*, one of the most prized shells among U.S. collectors. You can find it on Sanibel or Captiva after a storm, but usually just one half (the live animal lives far out at sea in depths up to 100 feet, so the shell tumbles for miles before washing up on the beach). Seven to nine ribs, covered with large, hollow knobs and bumps, make it look like a paw. A lion's paw can be dark red, dark orange or, rarely, yellow, and from 1 to 4 inches wide.

Facing page: Close up view of a stiff pen shell encrusted with barnacle shells, each less than a half-inch across

Right: Scotch bonnets

Man, I feel like a woman. Slipper shells live in stacks, up to eight high. When they mate, the sex organs of a male extend down until it finds a female. If there are no females, a male near the bottom will turn himself into one!

Atlantic slipper shell

Banded tulip

True tulip

Scotch bonnet

A **Scotch bonnet** *(Phalium granulatum)* looks like a pilgrim's bonnet from the 1600s. The shell has a fat, bulging shape, with a wide, flat shield next to the opening. Look for it after a storm.

Slipper shell

The inside of an **Atlantic slipper shell** *(Crepidula fornicata)* looks like a small slipper, or perhaps a small boat with a seat. In fact, slipper shells are also called boat shells, canoes and quarterdecks. If you find a well-shaped, balanced specimen, you can float your "boat" in the water (children love to do this). By the way, the piece you find on the beach is not a half shell, it's the whole thing. As gastropods, slippers have only one shell.

Tulip

The **banded tulip** *(Fasciolaria lilium)* is common here, and popular with collectors. Usually about 2 inches long — though it may be twice that — a banded tulip has a bulge in the middle, and two pointed ends. It's usually gray, but some have tan, green or orange tints or accents. The banded tulip is named for its thin, dark spiral lines, or bands, around its shell.

Larger than its banded cousin, the **true tulip** *(Fasciolaria tulipa)* can be 6 inches long. It has a darker color, sometimes nearly black, often with a green or red tint. Less common than a banded tulip, the true tulip can be found on Sanibel and Captiva beaches during the winter. Live shells will be tucked into the sand at low tide. Pick one up, and this aggressive animal will extend its foot and thrash about as it searches for a way to right itself.

Wentletrap

Because of its rib pattern, the tiny, delicate **angulate wentletrap** *(Epitonium angulatum)* is known as the Staircase Shell. In fact, the German word "wendel-treppe" means "spiral staircase." Gray or tan, the shell is ¼ inch to 1 inch long, with a large opening at the top. The shelling commu-

Angulate wentletraps are often in the debris at the high-tide line. Look at Lighthouse Beach.

nity recognizes Sanibel Island — particularly Lighthouse Beach — as the place in the world to find one. Look in the debris at the high-tide line.

Whelk

The **lightning whelk** *(Busycon contrarium)* is the only shell in Florida that spirals to the left. Streaks on the sides of younger shells are said to look like flashes of lightning. A live whelk feeds by prying open the shells of clams and other bivalves, sometimes breaking them open by whacking them against its own shell. Then it sticks its tongue — a bizarre tube with teeth on the end — into the prey to feed.

Lightning whelk

Worm shell

The tube-shaped **Florida worm shell** *(Vermicularia knorrii)* is a tube-shaped container that actually held a mollusk, not a worm. The coils are tightly wound at the base, then expand into wide, loose circles. It's also called an Old Maid's Curl. As a live animal, it moves about freely in the water. Related shells are the **irregular worm shell** *(Dendropoma irregularis)* and the **variable worm shell** *(Petaloconchus varians),* which live in mass colonies called "worm rocks."

Florida worm shell

DRAWING BY SAMANTHA BLAZE

'I Like to Find Shells in the Rain'

Whenever we see a painting of shellers doing the Sanibel Stoop — the iconic stance of bending over to inspect and pick up shells — it's always of older people. But shelling is fun for all ages, especially children. Here's what kids who live on Sanibel and Captiva have to say about shelling on the islands:

■ "The best beach for shelling is Gulfside City Park. Some of my favorite shells are the wentletrap, because it is the most beautiful shell, the horse conch, because it is our state shell, and the angel wing, because it makes you feel like an angel dropped her wing for you." — *Alicia Jordan, 14*

■ "I like coquinas because of their nice colors, clams because they open and close, and angel wings because they look like angels." — *Kacie Phillips, 9*

■ "Captiva has great shells. It is best to go at low tide." — *Kory Phillips, 11*

■ "Scallops have a lot of eyes." — *Matthew Wener, 8*

■ "You don't always find what you are looking for!" — *Emma Wheeler, 8*

■ "I love Bowman's Beach. The lace murex is beautiful. And the pear whelk is very light." — *Cherie Gelpi, 12*

■ "I like coquinas because I get to find them in the sand. Most people don't know that animals are usually still in the coquina shells." — *Micaela Neal, 7*

■ "Lion's paw shells are very lumpy, and angel wings are pretty. Shelling is fun!" — *Hannah Mitchell, 8*

■ "I like pretty scallops and clams." — *Ashley Swann, 7*

■ "I like to shell at the South Seas beach." — *Chelsea Grinstead, 9*

■ "I like to go to Bowman's Beach. Once I found a live Florida fighting conch. I like junonias, cat's paws and co-quinas." — *Taylor Bryant, 11*

■ "I like coquinas because they come in so many different colors. I also like lightning whelks and augers." — *Kirsten Harlow, 8*

■ "I like to find shells in the rain. My favorites are cat's paws and horse conchs." — *Samantha Blaze, 7 (as illustrated in her drawing above)*

Other sea life

Sand dollar

A round, flat animal, a sand dollar *(Mellita quinquiesper-forata)* is a type of sea urchin, an echinoderm. Designed like a spoked wheel, it has radial symmetry. Resembling a giant silver dollar made of sand, it makes a great souvenir.

A sand dollar has a five-point flower pattern on its top, with sets of small and large holes. The five large holes let food pass down to the mouth and help the animal burrow in the sand. Inside the "petals" are small slits where the sand dollar breathes. Eggs or sperm are released from the five small holes in the center, called gonopores.

These holes and markings have a religious significance to many people. The pattern on one side looks like an Easter lily. The other side looks like a poinsettia, a traditional Christmas flower. Legend says the large holes represent those made in the hands and feet of Christ. And the teeth — the hard pieces you find when you break a dead sand dollar open — are said to resemble tiny doves.

Dead sand dollars feel like sandpaper. Look for them in tide pools or just lying on the beach. Most are light tan or gray. (To make a sand dollar white, soak it in a 50-50 mix of bleach and water.)

Live sand dollars are brown and have bristles. Turn one over and you can see its tiny hairs, called cilia, which it uses to direct food to its mouth. Sometimes the cilia will be moving, in a wave-like pattern. A sand dollar lives

Facing page: Marin Williams, 7, collects sand dollars on Cayo Costa

You know you're a sheller when...

■ The back of your neck is the most sun-tanned part of your body.

■ You don't leave the beach when it rains.

■ You're just a little disappointed when the hurricane doesn't hit.

■ You travel to Sanibel with empty plastic bags, empty pill bottles and rubbing alcohol.

■ You stop for sand trails on the beach.

■ Friends think that you'll appreciate any gift with a shell theme, no matter how kitschy or tacky.

Sand dollars covered with tiny tusk shells

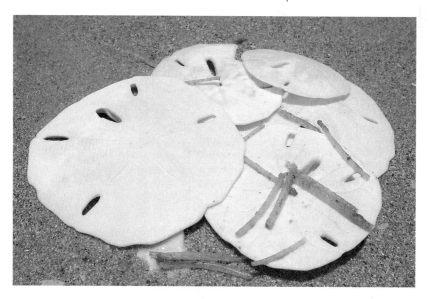

Sargassum weed

Top ten reasons to marry a shell collector

10. They love to go to nice tropical seashores.

9. They understand when you do something to make the house smell.

8. Shells look better than stamps on a shelf.

7. It's easy to decide what to give them for a birthday present.

6. They can't object when *you* spend money on *your* hobby.

5. Mating habits of mollusks give them ideas.

4. Less competition for the TV remote.

3. They'll go anywhere, if they can find shells.

2. Plenty of books for the coffee table.

1. Helps you learn Latin.

in the tidal zone of the beach — the area between high and low tide — as well as out at sea. It covers itself up in the sand, leaving only the top exposed. Unlike a dead sand dollar, which breaks apart easily, the live animal is solid and sturdy.

Sand dollars eat particles of food they find in the water. They, in turn, are eaten by starfish, snails and stingrays.

Sargassum weed

Out in the Atlantic Ocean, south of Bermuda, is the misnamed "Sargasso Sea." This huge mass of free-floating sargassum weed *(Sargassum natans)* is home to millions of small creatures, including crabs, shrimp and tiny fish — many of which have evolved to look exactly like parts of the weed, for camouflage. The whole thing is kept afloat by tiny air sacs on the leaves and stems.

Clumps of the yellow and brown seaweed break off, drift around Florida and wash up onto Sanibel and Captiva each winter. The animals come with them. Look closely at a strand and you may find a whole group of tiny marine creatures, sometimes still alive, that are completely different than any of our local sea life.

Columbus spotted the Sargasso Sea as he crossed the Atlantic in 1492. He thought he'd found land.

Sea anemone

The sea anemone *(Actiniaria sp.)* looks like a flower, with long tentacles that sway under the water. But to small fish it's a monster. When a fish makes contact, those tentacles sting and paralyze it. Then, like something from a 1950s horror film, the tentacles grab the helpless creature and bring it to the anemone's mouth for dinner. The mouth is located in the center of the tentacles.

Sea anemones have suction cups on their undersides, which they use to attach themselves to large shells and rocks. Look for one in tide pools and in gaps between rocks near the low-tide line. Pick one up and it may try to attach itself to you! Out of the water anemones turn into shapeless blobs.

Sea anemones

Sea horse

It has the head of a horse, the tail of a monkey, and the pouch of a kangaroo. And to top it off, the male gives birth! Is there anything stranger than a sea horse *(Hippocampus hudsonius)?* Only 3 or 4 inches long, a sea horse is a type of pipefish. But its face looks like a horse, with a long snout. Its body has prickly, spiny plates instead of scales.

A sea horse swims upright in the water, using its dorsal fin as a tail. It feeds by wrapping its tail around the stem of an underwater plant and sucking the plankton.

And, yes, the male has the babies. The eggs are created in the female, but she mates with a male by ejecting her eggs into him, in a slot below his stomach. The eggs are fertilized as they pass through the slot, then come to rest in a small inner pouch. A placenta-like material then develops within the male's pouch, allowing the eggs to de-

Shells have been put to use by man throughout the centuries. They've been used for art, jewelry, money, buttons, ink, road gravel, even chicken feed (they make the hen's eggs have stronger shells).

Sanibel was the first community in Florida to ban live shelling, in the late 1980s. The second, Fort Myers Beach, did so in 2000.

Sea horse

Sea pork

Sea urchin

Sea whip

Where you can't shell. An aquatic nursery, most of the bay side of the islands is part of the refuge. No collecting of any type is permitted there.

Facing page: Lexie Dekker and Julia Leal, both 7, show off a live sea urchin

velop. They hatch out about a week later, when the male goes through labor! He pushes out a few eggs at a time, through a series of spasms that can last for days.

Sea horses live just off the island beaches. At low tide, they'll be in the still water around the exposed sand flats, with their tails wrapped around pieces of seaweed. They wash up on the beach every now and then, after a storm.

Sea pork

For 20 years we've wondered what these things are. They've always looked like brains to us — the result of some gruesome mutiny at sea, perhaps. We've seen people afraid to touch one. We've learned they're called sea pork *(Ascidian sp.)*. Each of these harmless blobs of jelly is actually a large commune, full of hundreds of tiny sea creatures stuck together in a mass of solid gelatin. To see them, slice open a slab of sea pork and look at it closely in the bright sun.

Sea urchin

A sea urchin *(Lytechinus variegatus)* looks like a plant. But it's an animal, with five powerful teeth in its mouth, which is down on the underside of its body. Related to a starfish, a sea urchin is a small, round orb, small enough to hold in your hand. Short spines cover its body. You'll find sea urchins at the beach, in the water. Often one will be holding onto a rock or shell with its tube feet, to keep it from washing away as it eats passing seaweed. Some people eat sea urchin at sushi bars, where it's called uni ("OO-nee"). The only parts eaten are the sex organs.

Sea whip

This slender plant-like animal washes up on the islands after a storm. Also known as a soft or horny coral, a sea whip *(Leptogorgia virgulata)* has a flexible core, and perhaps a branch or two, covered with polyps. Dried by the sun it looks like a thin piece of wire, or whip. Most sea whips are red, yellow or purple. Sometimes you'll see small live shells, often Atlantic winged oysters, attached to one.

The World's Best Shell Collection
The Bailey-Matthews Shell Museum

The only shell museum in the world, the Bailey-Matthews Shell Museum *(3075 Sanibel-Captiva Rd., Sanibel; (888) 679-6450, 395-2233)* gives you a great introduction to mollusks, their shells and habitat, as well as the role they play in the environment. The well-organized museum has hundreds of interesting shells of every size, shape and color on display, most in educational exhibits. Plan on spending at least an hour here.

The focal point is the 72-foot-wide Great Hall of Shells, which houses a series of exhibits. The displays illustrate the role shells have played in ecology, medicine, literature, religion, art, architecture, and as a source of food. Exhibits include worldwide and local shells, with special attention to scallops, tree snails, cephalopods (octopus, squid and nautilus, including a 13-foot model of a giant squid), fossil shells and micromollusks

Actor Raymond Burr was a key museum benefactor

(shells as small as 0.5 mm). Kids love seeing the monstrous clamp-jawed shells.

A Sailors' Valentines display shows the unique shell boxes Caribbean women created for New England whalers to take home to their loved ones, an art form still practiced by some Sanibel residents today.

Dry aquariums and dioramas demonstrate local shell habitat. One 3-D exhibit captures a raccoon, fiddler crabs, and other creatures in the mangroves at low tide. The Florida land-shells display features a re-creation of Everglades habitat.

Daily slide programs explain the lives of mollusks, their habits and habitats.

The Children's Learning Lab has hands-on displays, a staffed live-shell touch tank, shell games and a video. Parental participation is encouraged.

The museum store has shell-related items ranging

from postcards to fireplace mantels and includes a nice selection of shell books and identification guides. The children's selection includes books, puzzles, games, stuffed animals, jewelry and clothing. For adult shoppers, the store offers books, china, hand-painted silk scarves and jackets, wind chimes, stationery, posters, garden supplies, jewelry, picture frames and more. There is no charge to visit the store.

Museum admission is $5 for adults, $3 for youth 8 to 16. Children 7 and under are free. Special group programs and tours are available at discounted prices. The museum is open every day but Monday. Operating hours are 10 a.m. to 4 p.m.

A collecting museum, Bailey-Matthews has over 200,000 shells. The collection is expanding, with a growing focus on mollusks from the coast of southwest Florida, its barrier islands, and the Gulf of Mexico. The in-

The children's lab has hands-on displays

creasingly regional nature of the collection has attracted national and international students and scientists, who visit to examine specimens in their fields of expertise.

The late actor Raymond Burr was among the founding fathers of the museum. Television personality Willard Scott has been another key contributor. Director Dr. José H. Leal was previously a scientist at the Muséum National d'Histoire Naturelle in Paris, France, and a postdoctoral fellow at the Smithsonian Institution in Washington, D.C. About 50,000 people a year visit the museum, which is named for two pioneer island families.

The Sailors' Valentines exhibit displays this unique Caribbean art form

Starfish

Starfish, also called sea stars, live in the shallow waters around Sanibel and Captiva. The **brown spiny sea star** *(Echinaster spinulosus)* is the one most people associate with the word "starfish." It has five thick, rounded arms coming out of a central disk. There are short, blunt spines on top, and hundreds of transparent sucker hoses (tube feet) underneath.

Brown spiny sea star

Look for the larger **nine-pointed star** *(Luida senegalensis)* living along the beaches, half buried in the sand. It has thin arms about 6 inches long. If you pick one up, don't be surprised if one of its arms "falls" off. They break off whenever the starfish feels threatened, and later grow back. You can also find **short-spined brittle stars** *(Ophioderma brevispinum),* just a few inches long, just off the shore or washed up on a beach after a storm.

A starfish feeds by using its arms to catch, and pull open, a shell; it especially loves oysters. It then turns its stomach inside out, pushes the stomach out through its mouth, and wraps it around the mollusk. The stomach digests the food right out there in the open. A typical starfish moves about 6 inches a minute. It can twist its body into many different positions, and creep through narrow openings.

More special finds

Crucifix shell

The skull of the **gafftopsail catfish** *(Bagre marinus),* known as the crucifix shell, symbolizes Christ on the Cross, including the spear wound. If you find one shake it to see if it rattles — a sign of extreme good luck. The rattling is said to symbolize the sound of soldiers throwing dice, gambling for Christ's garments (it's actually from two small bones inside which help the live catfish keep its equilibrium). Religious scholars say the gafftopsail, or sailcat, is the fish Christ used to feed the masses.

Dead man's fingers

The simplest of multi-celled animals, a sponge feeds by filtering food out of the water. When the **dead man's fingers** sponge *(Codium fragile)* feeds, tiny polyps emerge, giving its surface a fuzzy appearance. With its finger-like form, the sponge resembles a decomposing hand underwater.

Starfish can not only grow new arms, they can grow new starfish. Years ago starfish were the bane of the oyster industry. They would infest an oyster bed, eating every oyster in sight. To destroy the invaders, oystermen would catch the starfish and rip them up, tossing the pieces back into the sea. But each piece, if it contained part of the starfish's central disk, would then grow into its own, separate starfish.

Facing page: The "crucifix shell" is the skull of a catfish. It symbolizes Christ on the Cross.

Dead man's fingers

Parchment tube worm

Parchment tube worm

Those 5-inch-long sandy tubes you see on the beach are the casings of parchment tube worms *(Chaetopterus pergamentaceus)*. The live worm buries itself in wet sand in a "U" shape, with both of its ends sticking out. Water flows into one end of the tube and out the other; the worm feeds on the particles of food carried with the water. One or two pea-sized crabs *(Pinnixa chaeptopterana)* live in there, too. Like a firefly, a parchment tube worm will glow in the dark. If you're lucky enough to find a live worm on the beach at night, watch for its flashing blue light.

Lightning whelk egg case

Egg cases

How could this big long thing come out of a shell? Believe it or not, it does. **Lightning whelks** secrete their egg cases, a disk at a time, on the flats that surround Sanibel and Captiva. Many of these wash up on the beach. Each looks like a series of flat disks strung together. Female whelks string them out of their shells. The longer the string, the older, and larger, the whelk that made it. Each disk on the string incubates 20 to 100 baby whelks, each about the size of the head of a pin. When they hatch, they eat each other for food. Usually only one in each disk survives to grow into a mature whelk.

Ancient sailors called whelk egg cases "sea wash balls." They rubbed the cases together to produce a lather and wash their hands.

Clearnose skate egg case

The small, black purse-like containers on the beach are the egg cases of the **clearnose skate,** a relative of the stingray. These "mermaid's purses" average 3 inches long, about 1½ inches wide, and have a pair of inward-curving horns at each end. Hold one up against the sun and you may be able to see a single skate embryo inside. (Unlike skates, stingrays give birth to live young.)

The **Atlantic moon snail** lays its eggs into a collar-shaped gelatin "sand collar." Brittle pieces wash up on the beach. Active whole collars, full of eggs and tough and leathery, are common in the bay flats.

Tulip egg case

You can find egg cases from the **tulip shell**

attached to clam and pen shells. Laid in clumps, they look like small megaphones.

Shelling charters

A shelling charter will take you to nearby small islands and sand flats that are accessible only by water. There are often more shells here, especially rarer varieties, because there are less people collecting them. You'll also get a chance to see dolphins, birds and other wildlife. Call for reservations well in advance. Available charters include:

Up to four passengers can join **Capt. Mike Fuery,** author of "Capt. Mike Fuery's New Florida Shelling Guide," for a three-hour trip to the island of Cayo Costa and Johnson Shoals. Capt. Mike takes off from 'Tween Waters Marina on Captiva *(466-3649)* at 7:30 a.m.

Capt. Joe Burnsed, a Florida native and veteran guide, leaves from the Castaways Marina *(472-8658)* on the western tip of Sanibel. He heads to North Captiva and Cayo Costa. Capt. Joe also offers private lunch trips to Cabbage Key, Boca Grande and North Captiva. Both group rates and per person rates are available.

You can combine a shelling and fishing trip at **Memorable Charters.** Capt. Russ *(boat: 336-8377, home: 549-1426)* will take you fishing in Pine Island Sound, then shelling on Cayo Costa. All fishing tackle, bait and snorkel gear are provided. The boat departs from the dock at Punta Rassa Road, at the east end of the Sanibel Causeway. This charter is popular with families.

Sanibel Island Cruise Line *(472-5799)* offers shelling, snorkeling and sightseeing tours on larger boats, keeping per-person rates more affordable. The boats have private heads (restrooms), coolers and ice. Beach chairs, beach umbrellas, towels, snorkeling gear, shell nets and shell bags are also provided. Cruises last from two hours to all day.

What visitors say

Is Sanibel shelling really that good? We walked the beach one morning and spoke with shellers: "There are so many seashells on the beach! We weren't expecting so much." said Erin Smythe, a young mom from Idaho. "We took a charter trip to Cayo Costa and did some incredible shelling," said Bill and Patsy Johnston of Crystal River, Fla., a retired couple carrying bike helmets. Finally, we came upon Hilda Brandt, a 40-ish woman from Germany here with her husband. "Our friends came last year and said 'You've got to go there!' They were so right!" said Hilda. "We knew there would be shells, but not this many!"

The Song of the Sea. Put a big shell up to your ear and you'll hear the Song of the Sea. The sound of the surf has been mysteriously trapped inside the shell, to live forever!

That's the official line, anyway. The truth is a little less romantic.

The sound you hear inside a shell is the sound of the room you're standing in. The slight breeze in the room — nearly every room has at least a little movement of air — is shifting the air around inside the shell. This movement lets the shell amplify whatever small sounds are in the room — the hum of the air conditioner, for example — to create that "rushing wave" sound.

The shell creates its sound in much the same way the box of a guitar or violin does. The enclosure traps the air, which then bounces around inside, resonating whatever sound waves that get trapped in there. You can hear the same sound by holding a drinking glass up to your ear.

Listen to a shell in a closet, where the air is still and there are lots of clothes to muffle the sound, and you will hear almost nothing. But then take the same shell out to your living room, and — voilà! — the Song of the Sea magically returns.

Wildlife

*T*housands of interesting creatures make their homes on and around Sanibel and Captiva. Storks, spoonbills, stingrays, dolphins — they're all here, and relatively easy to see. Eric Carlsson, visiting from Sweden, says, "The wildlife here is incredible. I've seen many birds at the refuge, and many raccoons, even in the daytime." Adds New Hampshire's Tom Lamarck, here with his family: "We saw dolphins hunting for fish up on Captiva, and a huge family of stingrays at the beach. Then we saw spoonbills flying overhead by the school."

"Some of the days I'll always treasure are the hot but tranquil walks through Ding Darling with my binoculars," says Joanne Beagan, from the Jersey Shore. "I'm keeping a life list of bird sightings and the most exciting ones are from the refuge. My ornithological dream is to find the mangrove cuckoo. He's in there somewhere!"

"We told our travel agent we love birds, and she recommended Sanibel," adds Colleen Redmond, here from Hartford, Conn. "We haven't been disappointed!"

Birds

Sanibel and Captiva are the first major islands migrating birds come to as they fly north up the Florida west coast, making them a natural rest stop. Sanibel, in particular, attracts hundreds of unusual species, with its unique combination of freshwater and saltwater habitats. Serious birders — armed with a brain full of knowledge, decades of wisdom, and a $1,000 pair of binoculars — can spot up to 230 species of birds here, including stilts, cuckoos, hummingbirds and frigates. But our main attractions — the pelicans, herons, spoonbills and the like — are easy for anyone to find and enjoy.

Anhinga and Cormorant

The unusual **anhinga** (*Anhinga anhinga*) swims under the water to catch its food, then brings its prey to the surface and tosses it into the air to swallow. When swimming it sticks its head and long neck out of the water now and then to look around, earning it the nickname "snake bird." You'll see anhingas in brackish water throughout the islands. Look for them at the J.N. "Ding" Darling National Wildlife Refuge along the dike ditches by Wildlife Drive, especially near the culverts. Also look in tree branches a few feet above the water, where they spread their wings out wide to dry after fishing (an anhinga's feathers are not water-repellent, which allows the bird to dive through the water). Anhingas eat almost nothing but fish, sometimes snacking on frogs, snakes and baby alligators. *Aver-*

Facing page: A gopher tortoise eyes a snack near its burrow along Clam Bayou

U.S. FISH AND WILDLIFE SERVICE

Anhinga

Double-crested cormorant

Great egret

Facing page: Snowy egret

age length: 35 inches. *Average wingspan:* 45 inches. *Average weight:* 2.7 pounds. *Appearance:* Black body; yellow pointed bill; long neck; black legs. Males: Black head and neck. Females: Tan head and neck. Juvenile: All look like females until the third year.

The similar **double-crested cormorant** *(Phalacrocorax auritus)* also swims under the water to catch its food, and shares the same habitat. But it usually grabs, instead of spears, its fish. The "double-crested" name refers to two tufts of feathers that appear during the breeding season. *Average length:* 33 inches. *Average wingspan:* 52 inches. *Average weight:* 3.7 pounds. *Appearance:* Black duck-like body; orange, hooked bill and jowls.

Egrets

A **great egret** *(Ardea alba)* will walk right up to you. It has little fear of man. You can spot it in mudflats and marshes at low tide, hunting for fish and lizards, or beside roads, looking for snakes and insects. Some hang out at the beach, especially at Sanibel's Gulfside City Park. When feeding, the great egret will wave its head back and forth before striking — giving it stereoscopic vision to aid its aim — and will impale its prey on its long, sharp bill. When nesting, males will sit on the eggs for short periods of time to give their brides a break. The pair show affection by caressing each other with their heads. *Average length:* 39 inches. *Average wingspan:* 51 inches. *Average weight:* 1.9 pounds. *Appearance:* White, tall and slim; bright-yellow bill; S-shaped neck; black legs and feet.

The **snowy egret** *(Egretta thula)* finds food by sprinting about in shallow water, causing fish to dart out into the open. The bird rakes its toes slowly through the water, which also flushes out fish, insects and small vertebrates hiding in the sand. Some scientists believe the egret's yellow feet act as fishing lures. The distinctive feet have earned the bird the nickname "Golden Slippers." Male snowy egrets court females by extending their necks skyward, flying in a circle, then tumbling to the ground. *Average length:*

Great blue heron

Tricolored heron

**Yellow-crowned
night-heron**

24 inches. *Average wingspan:* 41 inches. *Average weight:* 13 ounces. *Appearance:* White, small and slim; black bill with yellow near the eyes; black legs; yellow feet.

The **reddish egret** *(Egretta rufescens)* performs a wobbly drunken-sailor routine when it fishes, lurching about with its wings spread. *Average length:* 30 inches. *Average wingspan:* 46 inches. *Average weight:* 1 pound. *Appearance:* Pale, slate-gray body; dusky, shaggy rose head and neck; gray legs and feet; pink bill with black tip. Juvenile: Pale chalky gray overall.

Herons

A **great blue heron** *(Ardea herodias)* is usually alone, hunting for fish and snakes. It stands as still as a statue, waiting for prey to come into its range. Then it strikes, snaring its meal, which can weigh up to a pound. You'll see the great blue in freshwater or saltwater, day or night. It flies with slow wing beats and has a deep, raspy call: "frahnk, frahnk, frahnk." Great blues protect their territory, sometimes fighting each other to the death. *Average length:* 46 inches, the tallest bird on the islands and the largest American heron. *Average wingspan:* 72 inches. *Average weight:* 5.3 pounds. *Appearance:* Slate-gray body tinged with blue; dark blue wingtips visible in flight; yellow bill; black plumes on head; pinkish-gray neck; gray legs.

The **tricolored heron** *(Egretta tricolor)* is the most abundant heron in the Southeastern U.S. It's easy to spot feeding on mud flats at low tide or perching on mangrove roots. *Average length:* 26 inches. *Average wingspan:* 36 inches. *Average weight:* 13 ounces. *Appearance:* Slender, bluish-gray body; white belly; white stripe down center of neck and chest; yellow bill; yellow legs. White plumes on head and blue bill during breeding season. Juvenile: Reddish head, neck and body.

The **yellow-crowned night-heron** *(Nyctanassa violacea)* munches on mangrove tree crabs the way kids gobble popcorn. Unlike other herons, it rarely eats fish. A stocky gray bird, it has a black head, a whitish-yellow crown and huge red eyes. *Average length:* 24 inches. *Average wingspan:* 42 inches. *Average weight:* 1.5 pounds. *Appearance:* Gray body; black and white head with pale-yellow plumes; gray bill; red eyes, yellow legs. Juvenile: Brown with white spots; no plumes.

The beautiful **green heron** *(Butorides virescens)* is a loner; it stays by itself as it crouches at the edge of the water to catch crabs. *Average length:* 18 inches. *Average wingspan:* 26 inches. *Average weight:* 7 ounces. *Appearance:* Small and stocky; dark, iridescent green back with gold markings; white undersides; greenish black crown; dark-brown neck; brown bill; orange legs.

You can spot the **little blue heron** *(Egretta caerulea)* feeding at low tide, often with a tricolored heron. *Average length:* 24 inches. *Average wingspan:* 40 inches. *Average weight:* 12 ounces. *Appearance:* Dark slate-blue body with purplish head; light-gray bill (blue during breeding season) tipped in black; pale-green legs. Juvenile: White; pale-gray bill.

Ibis

You'll see gangs of **white ibis** *(Eudocimus albus)* strolling across island roads and parking lots and at the beach. In the refuge they roost on the south side of Indigo Trail. Ibis feed by probing sand or mud with their bills, grunting as they hunt small fish, crabs,

Green heron

White ibis. Mature (above) and immature (below).

frogs, shrimp, insects and snakes. *Average length:* 25 inches. *Average wingspan:* 38 inches. *Average weight:* 2 pounds. *Appearance:* White body; red, downcurved bill; red legs; black-tipped wings. Juvenile: Mottled-brown and white body; paler legs and bill.

Osprey

An **osprey** *(Pandion haliaetus)* is a large fish hawk. It snatches fish from the water with its talons, then lands in a tree to eat its catch head first. Osprey mate for life. They prepare their nests in December and January, building their nests out of anything, including dolls, towels, hats and other found objects. A pair will use the same nest up to 30 years. Many stay on the islands year-round. Today there are 75 to 80 osprey nests on Sanibel and Captiva, up from 35 in 1979. *Average length:* 23 inches. *Average wingspan:* 63 inches. *Average weight:* 3.5 pounds. *Appearance:* Dark-brown body above with flecks of white; white chest and belly; distinctive dark eye stripe; gray feet. Often mistaken for a bald eagle. Females: Brown spots on chest.

Pelicans

The **brown pelican** *(Pelecanus occidentalis)* soars over the water, looking for a fish below. When it spots one it stalls, then dives straight down, crashing into the water at full speed, head-first, scooping up the fish in its pouch. Brown pelicans often fly together in long lines, sometimes just inches from the water. They flap their wings and then glide. The flaps are in sequence, starting with the leader. Each begins to flap when it reaches the spot where the leader began. Brown pelicans often coast just a few feet from the cars driving on the Sanibel Causeway. With your windows down you can hear their wings rustle. (Personally, a line of flying pelicans always reminds us of the flying monkeys in the Wizard of Oz. They had to be based on pelicans, right?) *Average length:* 51 inches. *Average wingspan:* 79 inches. *Average weight:* 8.2 pounds. *Appearance:* Distinctive pouch; grayish-brown body; dark-gray legs; orange and yellow bill; yellow cap in winter; dark-brown stripe on back of neck in breeding season (Dec.–Aug.). 1-year-old: Brown neck and wings, white belly. 2-year-old: Gray, lighter neck, darker stomach.

A large flock of **white pelicans** *(Pelecanus erythrorhynchos)* often gathers at the refuge,

Only a few days shy of leaving the nest, an osprey chick awaits its parents' return

When a pair of osprey is raising its young, only the male fishes. He will eat the head and tail of his catch, then bring the nutrient-rich middle section back to the wife and kids. Adults teach their young to fish.

Facing page: Adult brown pelican, with summer (breeding season) plumage, including dark brown neck stripe

Below: Left: 1-year-old brown pelican, with brown neck and wings. Right: Adult, with winter plumage and yellow cap.

White pelican

A roseate spoonbill is often mistaken for a flamingo

in an impoundment to your left about halfway down Wildlife Drive. Look for them about a hundred feet off the road. White pelicans fish by floating on the water and working together as a group, like a team of cattle ranchers. The birds circle around a school of fish, round them up in a herd, then scoop them up in their pouches. In a nest, the young put their bills into their parents' throats to eat partly digested fish. *Average length:* 62 inches. *Average wingspan:* 108 inches (the second largest wingspan of any bird in North America; the largest wingspan belongs to the California condor). *Average weight:* 16.4 pounds. *Appearance:* Distinctive pouch; white body; pink, yellow and orange bill; black-edged wings; orange legs.

Spoonbill

No, it's not a flamingo. It's a **roseate spoonbill** *(Ajaia ajaia),* actually a type of ibis. Look for spoonbills in the refuge, often across from the observation tower. Some roost in Clam Bayou. A spoonbill feeds by dipping its partly-opened bill in the water, swinging it from side to side, feeling for small fish, shellfish, insects and shrimp. It will feed in freshwater, brackish water and saltwater. Seeing spoonbills flying overhead is a stunning sight. Their large, pink wings light up the sky. A flock generally forms a diagonal line. Most spoonbills stay on the islands from March

LEE ISLAND COAST VISITOR & CONVENTION BUREAU

through September, spending the winter near the Everglades. *Average length:* 32 inches. *Average wingspan:* 50 inches. *Average weight:* 3.3 pounds. *Appearance:* Pale-pink to vivid-pink body; orange tail; gray featherless head; gray spatula-shaped bill; red eyes; red legs. Often mistaken for a flamingo. Juvenile: Pale, whitish body. *Note: Flamingos are seen on the islands only once every 5 to 10 years. Not native to the area, those that stop by are escapees from zoos.*

Stork

The **wood stork** *(Mycteria americana)* is the last surviving American stork. It lives only in a few spots in Florida, including Sanibel and Captiva. There are about 2,500 pairs in Florida, in less than 20 colonies. The wood stork was listed as an endangered species by the U.S. in 1984, because of dwindling numbers caused by national wetlands destruction.

A wood stork feeds by wading through the water with its beak open under the surface. When a fish touches it, the bill quickly snaps shut. Storks nest in the treetops of cypress or mangrove swamps. Parents keep their young cool by shading the newborns from the sun with their wings, and by carrying water in their throats and dribbling it over the chicks. Since a stork feeds by feel instead of by sight, its habitat has to be teeming with fish. A pair has to catch about 440 pounds of fish during a spring breeding season to feed themselves and their young. During a drought, the storks will travel to nest — less rain means less fish — settling as far away as New England or California. In flight, wood storks climb up to thousands of feet above the ground, and then soar for miles. As they descend, they dive, roll and turn.

Average length: 40 inches. *Average wingspan:* 61 inches. *Average weight:* 5.3 pounds. *Appearance:* White body with bald, blackish-gray head and black-edged wings; thick, downcurved, dark-gray bill; gray legs with yellow feet.

Shore birds

Many birds at the beach are best known as simply "sea gulls." The most common is the **ring-billed gull** *(Larus delawarensis)*, probably the most abundant gull in North America. *Average length:* 17.5 inches. *Average wingspan:* 48 inches. *Average weight:* 1.1 pounds. *Appearance:* White body with pale-gray wings; yellow bill with broad black ring; pale eyes; yellow legs. Juvenile: Mottled brown and white.

Overleaf: A wood stork feeds alongside Wildlife Drive. The closing of a stork's bill is one of the fastest movements in the animal kingdom.

Many birds associated with Sanibel nearly became extinct in the early 1900s. Using huge guns, mercenaries shot great and snowy egrets, spoonbills and ring-billed gulls by the thousands to sell their breeding feathers to the womens-hat trade. Stories are told of two men killing 300 egrets in a single day. Strict laws now protect these beautiful birds.

A secret birdwatching spot among islanders is the gazebo behind the Sanibel City Hall

Need binoculars? The Audubon Nature Store *(1985 Periwinkle Way, Sanibel; 395-2050)* sells quality brands. You can borrow a pair, free, at the refuge visitors center.

Ring-billed gull

'Hey, Isn't That a...'

Other birds to look for on the islands include eagles, hawks, vultures, woodpeckers, ducks, even a cuckoo.

The national symbol of the United States, the **bald eagle** *(Haliaeetus leucocephalus)* builds the largest nest of any bird in North America, up to 10 feet wide. There is a nest on Sanibel (on Silver Key, about a mile north of Bowman's Beach) and another on North Captiva. About a dozen bald eagles live on Pine Island and commute to Sanibel and Captiva to feed during the day. Florida has more bald eagles than any other state. *Average length:* 31 inches. *Average wingspan:* 80 inches. *Average weight:* 9.5 pounds. *Appearance:* Dark-brown body; solid-white head; yellow bill; yellow feet. Juvenile: All brown; head feathers whiten with age.

Look for **red-shouldered hawks** *(Buteo lineatus)* on dead trees. Seemingly faded from the sun, the hawks here are paler than those up north. *Average length:* 17 inches. *Average wingspan:* 40 inches. *Average weight:* 1.4 pounds. *Appearance:* Reddish chest and shoulders; pale-gray head and back, dark tail with narrow white bands; yellow legs. Juvenile: Streaky brown and white on back and wings; pale chest and belly.

Harassing a group of **turkey vultures** *(Cathartes aura)* could ruin your day — they just may throw up on you! As you might think, the vomit smells awful. Turkey vultures eat dead animals, either fresh or "aged," sticking their heads deep inside the carcass. They don't build nests, but rather lay their eggs beside shrubs and bushes. Look for them soaring overhead in a group, especially along Sanibel-Captiva Road, looking for roadkill or roosting together in dead trees. *Average length:* 26 inches. *Average wingspan:* 67 inches. *Average weight:* 4 pounds. *Appearance:* Large brown body with bold red head, like a turkey; silvery-edged feathers. Juvenile: Grayish head.

The cartoon character Woody Woodpecker was a **pileated** ("PILL-ee-ated") **woodpecker** *(Dryocopus pileatus),* the largest woodpecker in Florida. *Average length:* 16.5 inches. *Average wingspan:* 29 inches. *Average weight:* 10 ounces. *Appearance:* Black body with bright-red crested head; white under wings, stripes on face and neck; black eye stripe.

You'll often see a **common moorhen** *(Gallinula chloropus)* in a freshwater impoundment or roadside canal of the refuge. It runs over the top of the water when it takes off, creating a splashy wake. *Average length:* 14 inches. *Average wingspan:* 21 inches. *Average weight:* 11 ounces. *Appearance:* Small, dark blue-gray duck-like body; black head; red or orange bill. Juvenile: Drab gray bill.

The **American coot** *(Fulica americana)* is a slightly-larger cousin of the moorhen, and displays the same behavior. *Average length:* 15.5 inches. *Average wingspan:* 24 inches. *Average weight:* 1.4 pounds. *Appearance:* Small, dark gray-black duck-like body, black head; white bill. Juvenile: Dull, grayish-brown body.

The elusive **mangrove cuckoo** *(Coccyzus minor)* is here year-round, but few people ever see one. It the ultimate island bird to see for most birders. Look for this small bird sneaking through the mangroves. *Average length:* 12 inches. *Average wingspan:* 17 inches. *Average weight:* 2.3 ounces. *Appearance:* Medium-brown top; buffy belly; black tail with large white spots; black eye stripe; downcurved bill black on top, yellow below.

The smaller **laughing gull** *(Larus atricilla),* has a distinctive "laughing" call. *Average length:* 16.5 inches. *Average wingspan:* 40 inches. *Average weight:* 11 ounces. *Appearance:* White body with gray wings; black tail with white spots; black and reddish bill; black head in summer. Juvenile: Tan and white.

Laughing gull

The social **sandwich tern** *(Sterna sandvicensis)* gets better at fishing with age, as it dives from increasing heights and expands its hunting range. *Average length:* 15 inches. *Average wingspan:* 34 inches. *Average weight:* 7 ounces. *Appearance:* Similar to a gull. Mostly white with pale gray top and wings. Long black bill slightly tipped with pale yellow. Black eye stripe; black cap. Black legs and feet. Deeply forked white tail. Juvenile: Paler heads; tails less forked.

Sandwich tern

Willets *(Catoptrophorus semipalmatus)* also wander the beaches. They nest between April and June. Males and females take turns incubating their eggs, bowing to each other as they switch off. *Average length:* 15 inches. *Average wingspan:* 26 inches. *Average weight:* 8 ounces. *Appearance:* Large, stocky tan and white body with white belly; gray bill tipped with black; gray legs; striking black-and-white coloring on wings visible only in flight.

Willet

The 18-inch **American oystercatcher** *(Haematopus palliatus)* has a black head and a thick, bright-orange bill. The bill functions as a chisel, allowing the bird to pry open the shells of oysters and clams with ease. An oystercatcher will pretend its wing is broken to lure a predator away from its eggs. *Average length:* 17.5 inches. *Average wingspan:* 32 inches. *Average weight:* 1.4 pounds. *Appearance:* Brown back and wings; black head; white belly; bright-orange bill; black eyes rimmed with red and yellow; yellow legs.

Reptiles

Alligator

Floating in a bayou or basking on shore, an **American alligator** *(Alligator mississippiensis)* looks so still, slow and clumsy it's easy to forget it's essentially just a big lizard. Like a

American oystercatcher

Previous pages: An alligator rests along the bank of Indigo Trail

Even if you just see its head in the water, it's easy to figure out how long the whole alligator is. Just estimate the number of inches from its eyes to its nose, then convert this number to feet. Ten inches from the nose to the eyes means a 10-foot gator.

About 250 alligators roam Sanibel. Most are 6 to 10 feet long, and weigh about 250 pounds (the largest seen here was 14 feet long). There are few, if any, alligators on Captiva.

Where can you find an alligator? A Sanibel rule: If it's freshwater, there's an alligator in there somewhere. But they don't stay in one spot; alligators wander the island as they please. They can swim in saltwater, too.

Don't feed, touch or bother the alligators. An alligator accustomed to being fed associates humans with food, and loses its fear of man. The fine for alligator feeding is $500 and up to 60 days in jail. Gators that have been fed have to be killed. A Sanibel man was killed by an alligator in 2001, while walking his small dog.

lizard, it moves quickly. It can dart forward or sideways, or spin around, instantly. If you're standing to the side of an alligator, it can snap its tail and knock you over before you can blink. If you're in front of one, it can leap from 0 to 20 mph in just a couple of seconds, much faster than you can get away.

Alligators are an accepted part of life on Sanibel. Most are harmless to people if you keep your distance and don't feed them. In their natural state, they will usually remain still, or even crawl away from you. But if you provoke them, or they are protecting their nest, or they have been fed, alligators can attack. Even a tiny gator can bite and cause a serious wound, and if it's under a foot long the protective mother is sure to be lurking close by. To be safe, stay alert, stay respectful, and stay back at least 15 feet. Don't leave children or pets alone near any inland water on Sanibel, including hotel swimming pools.

An alligator eats only once or twice a week. It usually swallows prey whole — its 80 teeth are mainly for catching, not chewing (when a tooth falls out, another one grows in to replace it). An alligator hunts at night, eating fish, birds and turtles. If it can't find enough small prey, it goes after larger animals such as raccoons and bobcats. The gator drowns this prey underwater, then drags it up onto a bank to let it decompose for a few days, until it's soft enough to swallow — guarding it like a king's ransom until it's aged to perfection.

Sanibel alligators breed in April and May. The female builds a large nest, using sticks, mud and vegetation, and lays 30 to 50 eggs. Most babies don't survive — the adults love to eat them! And though alligators are among the most nurturing of the reptiles — the female cares for her young for nearly a year — even the mom will eat her offspring if they start competing with her for food.

Babies are about 6 inches long and have yellow stripes. When one does live past infancy, it grows about a foot a year for eight years, then about 6 inches a year after that. It can get up to 19 feet long, and live 40 or 50 years.

The skin of an alligator is tough. It's covered with smooth, horny scales, in rows connected to one another by narrow bands of thick, wrinkled skin. The back has rows of embedded bony plates. A gator's feet are webbed, like a duck. Pores in its head can sense even slight movements in the water. An alligator would be helpless without its tail. It's a weapon, a propeller, and a warehouse, where the alligator stores fat to get through the winter. The tail makes up half of a gator's length.

Since it's cold-blooded, an alligator will bask in the sun to stay warm.

Crocodile

An endangered species, only a couple hundred **American crocodiles** *(Crocodylus acutus)* are left in the U.S. But one lives on Sanibel. You'll usually find her in the refuge, often around Alligator Curve. You can tell the croc from a gator mainly by her size; she's almost 12 feet long. Also her snout is skinnier, and the fourth tooth on each side of her lower jaw is exposed when her mouth is shut.

This crocodile originally swam to Sanibel from nearby Pine Island years ago. The state of Florida trapped her and relocated her to swamps near Naples, about 30 miles south. But a year later she swam back to Sanibel. This time, officials decided to just let her be.

Every year she builds a nest and lays eggs. But since she has no mate, the eggs aren't fertilized and don't hatch. Female crocodiles (and alligators) can store sperm for up to 10 years, but this girl's been without a guy for decades. In 1995, she forcibly took over a nest of newborn alligators from their mother. She guarded the baby gators for about a week, responding to their cries. When she left the nest a week later — standard behavior in the croc world — the gator mom returned.

The croc hunts at night, eating fish and other animals that she finds around the water. She's about 40 years old.

She's quite welcome on the island. In fact, Sanibel has become an official Crocodile Sanctuary. If another croc comes here, it can stay, welcomed by island residents. Sanibel residents tried to start a crocodile breeding program, but the state of Florida vetoed the idea.

The Sanibel crocodile basks in the sun along Alligator Curve

Over a short distance an alligator or crocodile can run faster than a man. In the water it moves with explosive speed.

Secret wildlife
We asked island residents what wildlife they've seen on or around the islands, at least once, that visitors may not know to look for. The top 10 answers:

1. Nine-pointed starfish
2. Painted bunting
3. Octopus
4. Cottonmouth snake
5. Mangrove cuckoo
6. Kestrel
7. Whooping crane
8. Indigo snake
9. Batfish
10. Pink flamingo

Gopher tortoise

Tortoise crossings are a common sight on Sanibel

When you see a gopher tortoise or a turtle crossing a busy road, pick it up and help it across. Place it completely off the road, facing the direction it was going. But don't place a gopher tortoise in water. It can't swim.

You may see an indigo snake if you explore the islands' remote areas. It lives in palm hammocks, near ponds and in gopher tortoise burrows and stump holes.

Gopher tortoise

The **gopher tortoise** (*Gopherus polyphemus*) can't swim — it's not a turtle. It lives on land, eating plants. A "gopher," as locals call it, looks like a vintage military helmet, dome shaped, with four rugged legs sticking out of the sides. It averages 10 inches long, and weighs about 10 pounds. It can live up to 60 years in the wild.

Gopher tortoises live in burrows in the ground, usually 20 or 30 feet long and up to 8 feet deep, which they dig out with their front legs. They move from burrow to burrow, but stay within their territory of a few acres. The burrows provide homes for many other animals, too, such as snakes, armadillos, lizards, frogs and toads. There are tortoise burrows throughout the island, including in the refuge and along the trails at SCCF. Occasionally you'll see a tortoise walking along the side of Sanibel-Captiva Road.

The gopher tortoise is federally protected as a threatened species, and a "Species of Special Concern" in Florida. It's against the law to touch one, unless it's in danger.

Snakes

The **eastern indigo snake** (*Drymarchon corais*) is the largest nonvenomous snake in North America. It can grow to more than 8½ feet long. Dark blue, almost black, it is much stockier than the thin **southern black racer** (*Coluber constrictor priapis*), a smaller snake common on the islands. The indigo feeds on small mammals, such as palm rats, and especially on frogs, toads and other snakes. It eats its prey live, head first. Indigos have been classified as a threatened species by the Florida Game and Fresh Water Commission since 1971 and by the U.S. Fish and Wildlife Service since 1978. The main reason for their decline is habitat loss.

Four poisonous varieties of snakes roam through Sanibel and Captiva: the **eastern coral snake** (*Micrurus fulvius*), the **eastern diamondback rattlesnake** (*Crotalus adamanteus*), the **cottonmouth** (*Agkistrodon piscivorus*) and the **dusky pygmy rattlesnake** (*Sistrurus miliarius*). The coral snake is the most deadly. Island preschoolers learn the phrase "Red touches yellow, a dangerous fellow," to remember its unique red, black and yellow color scheme.

But the chances of being bitten by a poisonous snake on Sanibel or Captiva are almost nil. In fact, you'll rarely see one. Most snakes on the islands, such as black racers,

are not poisonous. Still, watch all snakes from a distance, and leave them alone. All species are protected, either by the city of Sanibel or the federal government.

Marine life

Dolphin

Bottlenose dolphins *(Tursiops truncatus)* are abundant in the waters surrounding Sanibel and Captiva. You can easily see them arc out of the water as they come up for air (a dolphin's fin comes up, out, and down; unlike a shark's, which stays parallel to the surface). Good spots are the causeway, the lighthouse fishing pier, anywhere along the beaches, or from the dock at Jensen's Marina on Captiva. Islanders head to the beach at dusk to watch groups of dolphins hunt for fish.

Dolphins are not fish. They're mammals; actually small whales. Like other mammals, a dolphin breathes air. It has a single nostril, called a blowhole, on the top of its head. The blowhole has a valve, like a plug, that stays shut until the dolphin comes to the surface. Dolphins usually come up to breathe about every 30 seconds, but they can stay underwater for up to 15 minutes.

All of a dolphin's sounds come out of its blowhole. The mouth is used only for eating.

A dolphin swims by moving its tail, called a fluke, up and down (fish move their tails left and right). To sleep, dolphins float in the water with their flukes hanging down

If you're on a boat you may spot a dolphin racing your bow, swimming behind your stern or jumping your wake (above). But watch closely. They often swim silently, without making the slightest ripple in the water, up to 22 mph.

As a dolphin swims just a few feet away, a lucky beachcomber at the Sanibel lighthouse gets a memorable photograph

Dolphins are incredibly smart. Using sonar — making high-pitched clicks and listening as the sounds bounce off nearby objects — a dolphin can find and identify objects that it can't see with its eyes. The dolphin can determine not only how far away an object is, but also what it is — a fish, a crab, a rock, another dolphin, anything down to the size of a BB — even if it's hidden behind another object.

Dolphins live in family groups, called pods, of up to a dozen. They swim, hunt, and care for each other as a group. When one gets sick, others guide it to the surface to breathe, and take turns towing it. Dolphins mate in the spring. A male courts a female by dancing around her, twisting his body into an "S." The female then decides if she wants the guy. Mates show affection by caressing and rubbing their snouts together. When the baby is born a year later, the pod gathers around the mother to protect her, then pushes the newborn to the surface for its first breath. Later, other females will pitch in as nannies and babysitters.

Facing page:
West Indian manatee

and turn around in a slow circle, still rising to the surface to breathe twice a minute. A dolphin can rest half its brain at a time, with only one eye closed.

Newborns are about 3 feet long, and weigh about 25 pounds. They can hear, see and swim immediately. Babies nurse about every 15 minutes — the nipples pop out of special slits in the mother's streamlined belly — and begin to eat fish in about six months. Around Sanibel and Captiva, babies are born between February and May.

A dolphin eats about 15 pounds of food a day, mostly squid and fish, with a bird thrown in for variety every now and then. Dolphins around the islands love mullet, a local fish. They find their prey at the surface of the water, and sometimes use their tails to injure or stun a fish before swallowing it. A dolphin has over 100 sharp teeth, so it can grind up about anything it wants. It swallows small fish whole.

A dolphin's two-tone color scheme (dark gray on top, nearly white underneath) makes it hard for fish to see it. A fish looking up from below is unlikely to spot the predator, as the white belly blends in with the water surface. A fish above can't see it either, as the dark back blends in with the dimly lit water below.

The natural enemy of a dolphin is a shark. But dolphins can swim faster than sharks, and sometimes will kill the attackers by whacking the sharks with their snouts.

Bottlenose dolphins can be up to 10 feet long, and weigh up to 650 pounds. They live about 20 years.

Manatee

To ancient sailors, the **West Indian manatee** (*Trichechus manatus*) looked like a beautiful, alluring woman, but with a tail. They called it a mermaid — half human, half fish.

But these sailors must have been out at sea a long, long time. A manatee looks like a huge, fat walrus.

The gray-brown mammal has paddle-like flippers and a round, flat tail (the Indian word for manatee means "big beaver"). Like a walrus, it has thick bristles on its upper lip. And it weighs a ton: an average of 2,000 pounds. Manatees are 9 to 13 feet long, and live 30 to 50 years.

Manatees live in the grass flats surrounding Sanibel and Captiva. The best place to see one is at a marina, especially on Captiva. In the refuge, it's not uncommon to see one from a small fishing boat or kayak.

Manatees come up to breathe every few minutes. They travel slowly, just a few miles per hour. Manatees move by flipping their tail flukes up and down, like dolphins. They use their flippers to move from side to side. They

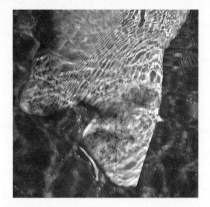

Going... Going...

Manatees are becoming extinct. They live in grass flats, usually in water less than five feet deep. When a fast-moving boat travels over these areas, the boater often cannot see a gray-brown manatee floating just beneath the surface, and the lumbering mammal has no chance to get out of the way. The hard, fiberglass bottom slams into the manatee and breaks its ribs, killing it. Even slow-moving boats seriously injure (and sometimes kill) the animals, as their propellers slice easily into a manatee's skin.

Among the manatees still alive in the bay, more than 90 percent have wounds or scars from boat propellers. The manatee in the photo above is missing the middle of its tail.

Florida is the only state ever to enact a law protecting manatees. As early as 1907, it was illegal to kill or harass a manatee here. The fine was up to $500 and six months in jail. In 1969, the manatee was officially listed as an endangered species. The Marine Mammals Protection Act of 1972 and the Endangered Species Act of 1973 strengthened manatee protection.

But the manatee population is still decreasing. More than 1,000 manatees have been killed by boat collisions since 1974. But some boaters, even locally, are still more concerned about their freedom to go where they want, when they want, than the manatee's right to survive. So manatees are still dying from broken backs and shredded bodies.

Poaching still occurs, too. Several men on neighboring Pine Island were arrested for selling manatee meat in 1996. Today a Pine Island man drives a pickup with a manatee license plate, meant to promote manatee protection. But his plate reads EAT-UMM.

Manatees are vulnerable to this abuse, as they reproduce slowly. A mother bears a calf only once every three to five years — the gestation period is 12 months and a baby stays with its mom up to two years. It's another three to five years before the calf reaches sexual maturity. Today there are about 3,200 manatees left, about 500 in Lee County.

Not everyone loves manatees. Some residents of neighboring Pine Island (above) resent the animals, arguing manatee zones unfairly restrict boaters.

can't see what's next to them, as their necks don't swivel.

Local manatees give birth to their young in Tarpon Bay and other secluded inlets. Once born, the mother immediately puts the baby on her back and lifts it to the surface. She then dunks the baby right back in the water, and lifts it up again. The mother does this over and over, until the baby establishes its own breathing pattern. Newborns swim with their flippers at first, learning to use their tail at about two months. Babies weigh about 45 pounds at birth, and nurse for about two years.

Manatees travel, though they usually stay in subtropical waters. Some locals have been seen as far north as Tampa. Manatees tagged on the east coast of Florida, at Jupiter, have shown up here. They appear to sense impending cold weather. Right before a cold front, the manatees around the islands swim over to Fort Myers to the warm waters of the Caloosahatchee River power plant.

Manatees eat the plants and grasses that grow in area bays and estuaries. They eat a lot, about 200 to 300 pounds of vegetation a day. Manatees really go for turtle grass, manatee grass and eel grass, all common in the shallow waters of Pine Island Sound.

Manatee fossils found in Florida date back 45 million years. Fossil skeletons in Jamaica suggest that manatees evolved from a four-legged, plant-eating land animal.

Sharks

Eleven species of sharks *(Carcharhinus sp.)*, including bull, hammerhead, blacktip, sandbar, nurse, tiger, bonnethead and blacknose, can be found around the islands. Amazing fish, sharks were around 100 million years before dinosaurs roamed the earth. A shark has no bones; its skeleton is made of cartilage, like a person's nose or ear. They have up to 3,000 teeth at a time, arranged in rows. When one falls out, another slides into place. Larger sharks, such as bulls and hammerheads, will fight fishermen in Boca Grande Pass for tarpon, sometimes taking the fish right off the anglers' hooks.

Bull sharks are the most likely to attack people, as they can swim in shallow water. But sightings from island beaches are rare. To stay safe, don't swim in the Gulf or bay late in the day, especially at dusk.

Stingrays

Related to sharks, **stingrays** *(Dasyatidae sp.)* are common in the summer along the island beaches and in the bays. This shy creature

Social animals, manatees show affection toward each other by rubbing their bodies together and smooching their noses and mouths together. Manatees, including mother and calf, also communicate by making sounds at each other.

Island kids tell us their favorite hard-to-find animals on Sanibel and Captiva are:
1. Hammerhead sharks
2. Octopus
3. Scorpions
4. Sea hares
5. Loggerhead turtles
6. Bald eagles
7. Indigo snakes
8. Squid
9. Reddish egrets
10. Coral snakes

To see the most wildlife, go out in the morning during a low tide. You can pick up a tide chart at the Sanibel Marina.

Sandbar sharks swim in the waters off Sanibel and Captiva. They come close to shore at dusk to feed.

IAN CARTWRIGHT

Southern stingrays hide in the sand as a southern and cow-nose ray glide by

To avoid getting stung by a stingray, do the Stingray Shuffle: As you walk in the water, scoot your feet in the sand, instead of taking high, deliberate steps. This will scare off any rays buried in front of you, and give you a safe path.

The stings are painful. If you do get stung you'll probably want to seek medical attention. You can get fixed up at Sanibel's HealthPark clinic *(1699 Periwinkle Way at Casa Ybel Rd.; 395-1414)* or the San-Cap Medical Center *(closest to Bowman's Beach and Captiva at 4301 Sanibel-Captiva Rd.; 472-0700)*. Most of the pain goes away in a few hours.

To care for the wound has fins attached to its sides in one smooth contour. It flaps these "wings" with slow, graceful beats to fly through the water. Its mouth is underneath its body; it feeds by sucking up small fish and other sea life.

Swarms of young southern stingrays routinely swim down the island shores in the summer, looking for food. American, cow-nose, yellow and spotted-eagle rays all roam the area. They swim just a few feet off the beach.

But when they rest they bury themselves in the sand, making them easy to step on. And that's how you get stung. A stingray is not aggressive, and it doesn't sting with the tip of its tail. It has a spike (or "spine"), sometimes more than one, at the base of its tail. When you step on its back you scare it, and it responds by flipping up its tail and zapping you with its spike. Ironically, the weight of your foot holds the ray still, letting it get a good jab at you.

Sea turtle

Loggerhead turtles *(Caretta caretta)* are air-breathing reptiles who live in oceans around the world, but always come to subtropical beaches, including those of Sanibel and Captiva, to lay their eggs. These large turtles can weigh up to 400 pounds, and be up to 4 feet long.

Loggerheads are tough to see, but their nests are easy to spot, since they are routinely marked and identified by volunteers. Look for these folks on the beach early each morning, April through October, starting at 6:30 a.m. They'll be happy to answer your questions about sea turtles and other beach wildlife.

Each sea turtle nest — a deep hole dug into the sand and then covered up again by the mother — contains up to 100 leathery eggs. (You can't touch the nests, but you can look at them; one of the best places to find nests is Bowman's Beach.) It takes up to three hours for a female to climb on shore, dig her nesting hole with her back flippers, lay her eggs, cover up the nest, and return to the sea. Females nest every two or three years, and make several nests during that season.

Each about the size of a ping-pong ball, the eggs hatch about two months later, usually at night. Immediately, the 2-inch hatchlings make a mad dash for the Gulf, using the light of the horizon to guide them there. It's a risky trip; the babies dehydrate if they don't reach the water within a few minutes, and birds and crabs love to snack on them as they crawl down the sand. Once in the sea, they're a food source for sharks and other sea creatures. Only about one in a hundred eggs will actually become an adult turtle, which takes at least 16 years.

At sea, loggerheads love to eat jellyfish and Portuguese men-of-war — the stings don't hurt them. But if they eat too many they get drunk: their eyes get puffy and red and they lose their coordination. Loggerheads are a threatened species, and are protected under state and federal law.

There were a record 536 sea turtle nests on Sanibel and Captiva in the 2000 season.

Horseshoe crab

The odd **horseshoe crab** (*Limulus polyphemus*) looks much the same as it did 500 million years ago. Up to 2 feet long and weighing up to 10 pounds, it has a hard outer shell — with a pair of scary-looking, though fake, eyes sculpted on the top — and a sharp spike of a tail. Underneath are five pair of legs and a couple of pincers. Look for horseshoe crabs throughout the wildlife refuge. Many hang out in the water along the east side of the Cross Dike.

Sometimes you'll see two swimming together, the smaller one on top. They're not mating, though that is the male on top. A male likes to attach himself to a female just before she crawls up on the beach to lay her eggs. He'll fertilize the eggs when she's finished, and is riding on her back to make sure he's the first male there.

You'll see shells of horseshoe crabs on the shore, near the water's edge or back in the bushes. They shed their shells as they grow, sometimes more than once a year. A horseshoe crab grows slowly; a 1-year-old is only the size of a nickel. It feeds at night, eating worms and mollusks along the sea bed.

yourself, first flush it out with water with Epsom salts, and check to make sure no part of the ray's spine is still in your foot. Soak the wound in hot water for an hour, or until the intense pain stops.

You must see a doctor if any part of the ray, or any other foreign matter, stays in the wound; if the cut is big enough for stitches; or if you feel you need an anti-tetanus shot. On rare occasions someone is stung in the stomach or chest. These stings are very serious, and require immediate, professional medical attention.

Female loggerheads return to the beach of their birth to lay eggs

Adopt a sea turtle nest. The Sanibel-Captiva Conservation Foundation's "Adopt-a-Nest" program gives you a chance to adopt a loggerhead turtle nest of your own. SCCF will monitor your nest and tell you how many turtles hatched from it. You also get a T-shirt and a free subscription to a turtle newsletter. For details call SCCF at 472-2329.

A horseshoe crab crawls up onto the sand

You can be stung by a moon jellyfish (below) while swimming in the water or by picking one up off the beach. Like a bee sting, it's painful, and it can produce an itchy rash. Use meat tenderizer and ammonia to relieve the pain.

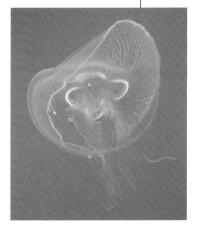

Out in the Gulf a horseshoe crab sometimes swims at the surface upside down, paddling with its legs. And though the tail looks like a weapon, it's simply a rudder.

Not really crabs at all, horseshoe crabs are related to spiders and scorpions. They are the only surviving members of a large group of animals that first appeared millions of years ago, before dinosaurs.

Jellyfish

The **moon jellyfish** *(Aurelia aurita)* is the most common jellyfish here. It has a round body with a fringe of tentacles, and can be up to 16 inches wide. It swims gracefully through the water, as if in a ballet. You'll see moon jellyfish on the beaches after a storm, or at times swimming in the water. Not actually fish (or jelly, for that matter) jellyfish don't have a brain, heart, blood or gills. In fact, they're 95 percent water. They do have sex organs, though. The male's are pink, the female's brown.

Crabs

You'll find the **blue crab** *(Callinectes sapidus)* in brackish areas such as around the culverts on Wildlife Drive. Actually greenish-blue, it average 6 inches wide and 3 inches long. Crabbing is allowed in the refuge, but throw the females back; only one ten-thousandth of one percent of the eggs survive to become adults. You can determine the sex by the shape of the crab's abdomen: the female's is curved; the male's more of a rectangle.

You'll spot the **fiddler crab** *(Uca minax)* in mangrove habitats, down in the muck by the roots, living in small holes it digs in the sand. When it crawls out, it meticulously pushes the sand out in front of it with its pincers. Males have one claw that's much bigger than the other; they use the big one to attract females and protect territory. Don't be surprised if a male waves his big claw at you.

Look for the **hermit crab** *(Pagurus sp.)* at the beach. A hermit crab has no armor on its tail, so it protects itself by living in a snail shell. As it grows, it finds bigger and bigger shells to call home. If it needs to, a hermit crab will pull a live snail out of its shell, eat it, then move in.

Also at the beach you'll find the **sand flea** *(Emerita talpoida),* also called the mole crab. This 1-inch-long creature burrows in the sand like a crab, but backward. Its antennae stick together to form feeding tubes, which you can see sticking out of the sand. You'll find sand fleas right where the waves are breaking on the shore.

As its name suggests, the **mangrove tree crab** *(Aratus pisoni)* lives among the mangrove trees, eating red mangrove leaves and debris. About 3 inches long, the slow-moving brown crab looks like a knot on the tree. Look closely to see the sharp tips on its legs, which allow it to climb trunks and roots. The best place on the islands to see it is at the Red Mangrove Overlook in the refuge.

The **angulate periwinkle** *(Littorina angulifera)* tree snail lives on the roots and branches of mangrove trees.

We asked island visitors what wildlife they enjoyed seeing the most:

■ **Kids under 6** loved seeing stingrays and the bigger wading birds. If you've got kids this age, a hike up Bowman's Beach or a walk down Indigo Trail might fit the bill.

■ **Elementary kids** were partial to alligators, dolphins and raccoons — maybe a refuge hike and a day out on the water.

■ **Teens** loved gators, dolphins and sea turtles. SCCF offers summer sea turtle programs.

■ **Adults 18-54** loved seeing dolphins and osprey. That's a day in the bay, another at the refuge.

■ **Seniors** were bird freaks. So maybe just the refuge itself, at low tide.

Angulate periwinkle

"It's a panther!" Each year about a dozen people call the refuge to report seeing a Florida panther roaming the islands. But they're seeing bobcats. The Florida panther, a type of cougar, is much bigger than a bobcat — it's up to 6 feet long and weighs up to 200 pounds (just its tail can be longer than a bobcat). There has never been a confirmed panther sighting on the islands, and no one has found panther tracks or claw marks here. One reason panthers are not on the islands is that the animals they feed on, deer and wild hogs, are not here, either. Though only 60 to 70 Florida panthers exist in the wild, most do live in southern Florida.

A raccoon track along the trails at SCCF

Facing page: A monarch butterfly rests in the butterfly house at the Sanibel-Captiva Conservation Foundation

Just an inch long, it climbs up the trees at high tide, scraping lichens off the branches for food, and climbs back down into the mud at low tide. It spends most of its life dry, but lay its eggs into the water. Often it has a pale periwinkle-blue tip. A good place to spot an angulate periwinkle is at the Red Mangrove Overlook.

Other animals you may see

Bobcat

Your best chance to see a **bobcat** *(Felis rufus)* is right at dusk or dawn. Often looking like a dirty, overgrown house cat, the 2- or 3-foot-long bobcat is fairly common here. We've seen them near Gulfside City Park, in subdivisions at the west end of the refuge, and walking along Sanibel-Captiva Road.

You can tell a bobcat from a domestic cat by its larger size, pointed canine teeth, bigger ears and "bobbed" tail. A bobcat also has a spotted coat, though the spots are often covered up by sand or other debris.

A bobcat sleeps during the day (in its den, often built in a hollow log) and hunts at night. It stalks its prey, pounces on it (leaping up to 10 feet), and often kills it with a single, powerful bite. Some islanders have lost their pets this way. As you might expect, bobcats are dangerous when cornered or confined.

Raccoon

Raccoons *(Procyon lotor)* thrive on the islands. Although nocturnal, these large, masked rodents roam the islands day and night, often in families. To see one easily, look around a restaurant dumpster after dinner — raccoons love the ones at the Lazy Flamingo restaurant on Periwinkle Way. Drive down Sanibel-Captiva Road at night and you're almost guaranteed to see a few raccoons (be careful: they often run across the road). Though they look adorable, raccoons draw mixed reactions from island residents. "They poop all over people's cars," says Gigi Claiborne, a student at the Sanibel School. *Tip: Don't leave food in an open car on the islands, especially at night. Raccoons will climb in and eat it while you're away.*

Also on the islands

An endangered species (it was nearly hunted to extinction for its pelt in the 19th and 20th centuries), the **river otter** *(Lutra canadensis)* lives happily here. Ironically, the best place to see one is in the sea, behind the Green Flash Res-

When you see someone else looking for birds, join in. Birders are usually friendly and helpful, willing to share their knowledge. We were all beginners once. If all else fails, just go to the refuge with your binoculars. Someone is sure to strike up a conversation and they might lead you to a whole new group of birding buddies.

Caged animals. Sanibel offers two spots to see wildlife from other areas:

■ Exotic birds, miniature deer, lemurs and monkeys are at the Periwinkle Park Campground *(1119 Periwinkle Way, Sanibel; 472-1433)*. Camping is for paid guests only, but the attendant will let you in to look at the animals (above) free of charge if you ask nicely. Children love it here.

■ The campground has some of its birds in the courtyard outside Jerry's Foods *(1700 Periwinkle Way, Sanibel; 472-9300)*. Cages hold macaws, cockatoos and parrots. These birds actually talk — when they feel like it.

taurant on Captiva. This freshwater weasel can swim in saltwater and loves to play around the docks there.

About the size of a cat, an **opossum** *(Didelphis virginiana)* is easy to spot at night along Sanibel-Captiva Road. The opossum carries its young in its pouch, like kangaroos and other marsupials. Keep an eye out for these guys as you drive, too. They move slowly, and aren't too bright.

Covered with bands of armor, the **nine-banded armadillo** *(Dasypus novemcinctus)* rolls up into a ball when frightened. It can be tough to see this shy creature on the islands, but many are out there. Listen for its squeak.

Look for **marsh rabbits** *(Sylvilagus palustris)* at dawn or dusk in the Bailey Tract. These small rabbits walk instead of hop, and have a tiny brown tail. Sanibel's Rabbit Road was named for the abundance of marsh rabbits there.

Florida's state butterfly, the **zebra longwing** *(Heliconius charitonius)* is black with yellow stripes and spots. The adult has a wingspan of up to 4 inches. As a caterpillar, it's white, with long black spines and a pale yellow head. Just before it splits open, the chrysalis is almost transparent. Look for them in the refuge or on the trails at SCCF.

Be careful when picking up dead palm fronds — **brown scorpions** *(Centruroides gracilis)* love to hide on them. You'll also find brown scorpions under boards or fallen tree branches. Island scorpions are 1 to 6 inches long. The sting from a brown scorpion can make you sick, but it won't kill you; it's not as poisonous as other varieties. Ammonia will help ease the pain.

Pests

Saltwater mosquitoes *(Aedes australis)* breed so profusely here that you'll often see low-flying helicopters, and sometimes even a DC-3, spraying the islands with repellent. In 1953, before mosquito control, Sanibel set the world's record for the most mosquitoes caught in a light trap in a single night: 365,000. To avoid attracting mosquitoes, don't wear perfume and cologne.

Sand flies *(Ceretopogonidae leptoconops)* are so tiny they are almost invisible; islanders call them "no-see-ums." They cut your skin and lap up the blood, and seem to have an insatiable fondness for visitors. When we used to come here on vacation we would get covered with no-see-um bites; we still have the scars to prove it! To keep sand flies off you, try a repellent that contains less than 20 percent DEET. Many island residents use Avon's Skin-So-Soft.

Fire ants *(Solenopsis invicta)* are aggressive, and their bites hurt. Step on a fire ant mound — a pile of sand about 6 to 12 inches high, often at the side of a road —

and these red monsters will attack your foot and leg with a vengeance. The ants are less than ⅛-inch long, but the bites will swell and itch for days. Not a native species, fire ants were brought into the United States by accident, when a boat from Brazil docked in Alabama in the 1920s. They spread to southern Florida in the early 1970s.

Where to see wildlife

For birds and alligators, the place to go is the **J.N. "Ding" Darling National Wildlife Refuge** *(off Sanibel-Captiva Rd., 2½ miles west of Tarpon Bay Rd.; One Wildlife Dr., Sanibel; 472-1100),* a 6,400-acre preserve that takes up nearly a third of Sanibel. Created as a migratory bird sanctuary, the refuge is one of the best birding spots in the U.S. Over 230 species have been seen here. During the winter, come at low tide to see hundreds of subtropical wading birds and migrating waterfowl, as well as alligators, otters, raccoons and an American crocodile. Even in the summer you can find plenty of animals. Roseate spoonbills are here year-round.

Volunteers are often out in the refuge ready to answer questions or let you view an animal through a spotting scope. Look for them along Wildlife Drive in golf carts.

Drive down the five-mile-long Wildlife Drive, newly paved in 2001, for easy access to several great birding spots. Walk along the

Birders find plenty to see along Wildlife Drive

Scan Sanibel-Captiva Road as you drive, watching the edges for wildlife about to cross. The road separates two rich island habitats — the saltwater and brackish refuge on one side, the freshwater SCCF land on the other — so animals cross back and forth a lot, especially at night.

Refuge tram

Bird guide Bev Postmus prepares her guests for an SCCF birding tour

Help our wildlife by not littering. Any food wrapper, apple core or soda can tossed out of a car will attract raccoons and other animals to the side of the road, often with fatal results.

Which species do birders want to see most? Guide Bev Postmus says roseate spoonbills are at the top of the list. Anhingas are a favorite, too. In the summer, birders search for a mangrove cuckoo, or nesting birds such as a black-whiskered vireo, black-necked stilt, snowy plover or gray kingbird. (Gray kingbirds nest in the Bailey's General Store parking lot.)

shores of the Cross Dike to find herons, anhingas and fiddler crabs. Stop at the Red Mangrove Overlook to see mangrove tree crabs and angulate periwinkles. Hike down the sandy, two-mile Indigo Trail to experience more remote areas. On the left ibis roost and feed, on the right look for spoonbills. Alligators routinely cross this path; look for their tail-and-claw prints crossing from one side to the other. If it's your first time to the refuge take the guided tram tour first. It's informative and relaxing.

The Bailey Tract, a separate piece of land off Tarpon Bay Road, has two miles of trails through freshwater habitat. Look for birds, raccoons, bobcats, alligators and snakes.

Four miles of trails lead through pristine habitat at the **Sanibel-Captiva Conservation Foundation** *(3333 Sanibel-Captiva Rd., Sanibel; 472-2329)*. Alligators, wading birds and other wildlife feed here. You can roam around by yourself, take a guided tour or join in on a birding trip. A butterfly house is also outside. Inside there's a touch tank, as well as taxidermy osprey, bobcats and other local animals. Admission is $3 for adults, free for children under 17. In the winter the foundation has guided tours of Sanibel's beaches and marinas on Thursday mornings, giving you the chance to spot dolphins, manatees and shore birds.

For migrating birds also try the **lighthouse area** from mid-April to mid-May. For shore birds, dolphins and other sea life any of the beaches are terrific. Rent a boat to search for dolphins yourself out in the bay. Manatees are most likely to be seen at the South Seas Marina or 'Tween Waters Marina. The Sanibel Causeway is a great viewing spot for pelicans and dolphins.

You rarely see another person on **Clam Bayou,** a Gulf-side estuary at the west end of Sanibel. Silver Key, an island within the bayou a mile north of Bowman's Beach park, has a small trail and a bald eagle nest. Launch your canoe or kayak at the small dock off the park's spillover-parking lot. Much of this area needs restoration.

Guided tours

You'll join the former mayor of Sanibel at **Canoe Adventures and Wilderness Tours** *(472-5218)*. Mark "Bird" Westall will take you on a canoe trip through the hidden spots of the wildlife refuge, down the Sanibel River or to Buck Key, a small island off Captiva. River trips can be at night. One of the islands' true wildlife experts, Bird shares his wisdom as well as his opinions on mankind's environmental responsibilities. He keeps his trips small, to just a few people at a time. He has two 17-foot canoes that hold three people and one 20-footer that fits five. Trips are available any day but Friday.

Birding tours can be for beginners too. Local trips are arranged by the Sanibel-Captiva Conservation Foundation *(3333 Sanibel-Captiva Rd., Sanibel; 472-2329)*, often on Friday mornings.

Learn more about island birds at the Sanibel Public Library *(770 Dunlop Rd, Sanibel; 472-2483)*. It has copies of Birder's World and Bird Watcher's Digest magazines, as well as the latest issues of the Pileated Woodpecker newsletter of the Sanibel-Captiva Audubon Society and the Osprey Observer newsletter of the Sanibel-based International Osprey Foundation.

CROW

Fans of Animal Planet's "Wildlife Emergency" will recognize the **Clinic for the Rehabilitation of Wildlife** *(3883 Sanibel-Captiva Rd., Sanibel; 472-3644)*. CROW has been featured in 11 episodes of the popular television program. The wildlife hospital and rehabilitation center cares for animals from throughout Southwest Florida. Over 2,000 patients are treated annually. Ninety percent have been injured by humans, usually by being hit by a car or caught up in fishing line or hooks. Once an animal recovers it is moved outside to get used to the natural world again.

The clinic is open 8 a.m. to 5 p.m. You can drop off patients until 8 p.m., seven days a week (after hours leave them under the building for safety from predators).

Funded by community donations and grants, CROW gets no government money. Volunteers make up a huge proportion of the staff. The two Sanibel supermarkets, Bailey's General Store and Jerry's Foods, contribute food. Baby birds eat scrambled eggs, softened cat food and smashed citrus fruit.

CROW presents a short lecture, video presentation and small tour of its outdoor cages most weekdays at 11 a.m. and most Sundays at 1 p.m. The cost is $5 (free for children). Call in advance to confirm tour days.

Below: Looking into the rehabilitation cages at CROW. **Bottom:** A pelican recuperates.

Would 'Flipper' Bite a Human?

By Mark 'Bird' Westall

A few years back, the Duchess of York was on national network television promoting ecotourism. Great! We need celebrities championing the concept of coexisting with our natural environment. The problem is, the way she did this was by participating in a scuba "ecotour" in the Bahamas, where they gave her chain-mail to put over her wetsuit so she could go down and hand-feed the sharks!

Now first of all, this tells me how bored we all are in our society, that we have to get our jollies by hand-feeding sharks. But there may be more serious consequences to this type of behavior. Scientists in Florida can't understand why shark attacks are on the increase along the east coast of the state. Well, duh. I don't think that it takes a genius to suggest that we have a lot of very tame sharks swimming along the Gulf Stream between the Bahamas and Florida that used to basically ignore swimmers (along the Florida coastline), but now are approaching more of them, looking for handouts.

The same thing is happening with dolphins throughout coastal Florida. I had a fishing guide here on Sanibel tell me that he makes bigger tips if he feeds the dolphins on his trips and brings them in close to his boat. But guess what? The dolphins also are becoming more aggressive and biting more and more swimmers in the areas where these dolphin hand-feeding tours are occurring.

And I love the reaction of people when I tell them about the dangerous dolphins just after I've told the story about the increased shark attacks. They can easily accept the idea that a shark would want to attack us, but when they hear about the dolphins, their usual response is, "What? Flipper bite a human? Flipper wouldn't do that! Flipper's our friend!"

Flipper isn't biting those people to eat them — he's biting them to get their attention. And when a large animal like a dolphin or a shark bites someone, they bleed. The problem is, we tend to blame the animal and not the people who created the problem.

As a matter of fact, I firmly agree with much of the new research that tends to show that most shark attacks are accidents. Sharks are nearsighted and think they are hitting prey other than humans. That's why so many people survive shark attacks and only receive one initial bite. Basically, we may not taste good to a shark. However, once any large carnivore — such as a shark, dolphin, grizzly or an alligator — gets put on animal welfare, it begins to approach people looking for free handouts; and it literally doesn't know the difference between the hand (or leg) and the handout.

So how come hand-feeding is such a bad idea, if it makes so many people feel closer to nature? The answer is simple, really. Food is the most powerful stimulus to all living creatures, including people. An animal which has been hand-fed by someone naturally becomes more aggressive towards people because its own selfish genes alter its natural fear, or at least respect, of humans and it begins to act as though the reason that ALL humans exist is to give them food for free. Why go out and hunt for a living, if you can just hang out around humans and get a free lunch?

Understanding these basic concepts, in 1976 Sanibel was the first community in the state of Florida to make it against the law to feed alligators. Naturalists living on the island, along with the refuge staff, recognized the increasing conflicts which were surfacing in Yellowstone National Park with grizzlies being fed by the tourists. They postulated that this was analogous to our alligator situation here on Sanibel. Eventually, the state followed Sanibel's lead and it is now against state law to feed alligators. But even though it is against the law to feed alligators everywhere in Florida, it is amazing to me how many people have found the urge to feed these potentially dangerous animals irresistible.

When I used to live near the exit of Wildlife Drive, the alligators would sun themselves along the bank which paralleled the Drive and my backyard. I could sit inside my house when the windows were open and hear people say, "Oh, look dear. He's eaten the cracker already!" At first, I would run out and yell for them to stop feeding the alligator because it was against the law. The response I got was, "So what? Everybody breaks the law." The next time it happened, I ran out and said, "Don't feed the 'gator because it will become more aggressive towards people and we'll have to have the animal destroyed." The reply I got back was, "So what? You have lots of alligators on Sanibel."

Years later, shortly after a little girl was killed by a tame alligator up in Englewood, Fla., I put up a sign in the backyard which stated, "*Please help protect our children!* Don't feed or throw *anything* at the 'gators. Your actions make them more dangerous towards humans and that threatens our children!" Isn't it amazing what psychological games have to be played to get our "intelligent" species to understand the consequences of our actions?

Concurrently with the death of the Englewood child and the placement of my sign, the refuge staff did a little study of their own trying to determine if people were negatively interacting with the alligators along the Alligator Curve of the Wildlife Drive. During a two-week study, volunteers observed six people feeding the alligators and 20 people throwing rocks or shells in the direction of the alligators to get them to move.

The throwing of the rocks and shells is an example of random reinforcement. If you want an animal to do a trick, don't reward him with food every time he does what you want him to. That way, he will do the asked-for behavior whether he gets a reward or not.

To an alligator, the rock or shell hitting the water imitates the sound of food hitting the water, and the gator just assumes that he missed the food that time. Eventually, all a person has to do is approach the edge of the bank and the alligator moves toward the person in anticipation of the handout. The scary part is: what if the person is only a small child? It is a tragedy waiting to happen.

But what really scared the refuge staff during the study is that six people walked over to the alligators sunning along the edge of the road and actually *touched* the alligators. Now, I think we should live in a free society. If people want to do risky things like climb sheer cliffs or bungee-jump, they should be allowed to (as long as I don't have to pay their hospital or funeral bills). So if a person wants to touch an alligator… hey, it's a free country. But if that alligator grabs hold of that person and pulls him down into the water and drowns him… well to me that is natural selection at work! That human was not a very bright creature and, hopefully, hasn't bred yet, so he won't be passing on any genes that say it is OK to go over and touch alligators. The problem is, we live in a society that says, "No, that person has a *right* to be stupid. And if he's going to be that stupid, well then, we need to remove the alligator from the area so that individual can continue his dimwitted ways."

One incident observed during the study was a father who actually took his toddling child over to the side of a sunning gator and had his child *pet the back* of that lounging reptile. How do we protect our children from the actions of irresponsible adults? Do we destroy the alligators to remove the risks?

The feeding of wildlife, particularly of alligators, is a constant concern of the refuge staff. The next time you drive through Alligator Curve, notice that there are no longer any basking areas within reach of people. Vegetation buffers have been planted to limit the risk, but the feeding problems still persist.

We as a species seem to have an irresistible urge to feed animals.

— *A former mayor of Sanibel, Westall today runs Canoe Adventures and Wilderness Tours. He lives on the Sanibel River.*

Indigo Trail

White Ibis

Hikers and bikers are welcome, but for your protection and the protection of this fragile ecosystem, please stay on the trail.

Gulf of Mexico

Cross Dike = 2 Miles

daily. Closed at the Cross

Mopeds/vehi
Collecting
Feeding or disturbing
Pets not on a leash
Entry into a Closed Area
Boating
Crabbing with baited line or T
Possession of firearms
Littering
Camping
All violators subj

can Alligator

r nothing bu
ave nothin
footpr

BEAC

Refuge

*T*he pleasures of Sanibel's J.N. "Ding" Darling National Wildlife Refuge are subtle. This is not a zoo or theme park. In fact, it's not a park at all. It's a refuge for wild animals, home to large exotic birds and even larger reptiles, a combination of subtropical forests, marshes and hardwood hammocks. Two hundred and thirty species of birds can be found here, as well as 50 species of reptiles and amphibians and 32 species of mammals. The animals are never fed by man, are not fenced in, and come and go as they please. But their habitat is protected with a vengeance.

Ironically, this emphasis on animals over humans makes the refuge incredibly popular with people. A million a year come here, more than come to any other refuge in the United States.

Wildlife is easy to see. Especially birds — "Ding" Darling is the best birding spot in the United States. The prime season for birding is the winter, when migratory species fill the estuaries. The five-mile Wildlife Drive meanders through many idyllic habitats, letting you drive, bike or walk within a few feet of exotic creatures. Miles of hiking and biking trails take you through remote feeding and nesting grounds. There are water trails, too, and kayaks to explore them.

The 6,400-acre refuge is actually several individual tracts of land. The largest is the **Darling Tract,** 4,900 acres of mangrove wetlands and mud flats on the bay side of Sanibel. Flocks of migratory and wading birds feed here, especially in two brackish impoundments totaling 850 acres. The entrance to the Darling Tract is on Sanibel-Captiva Road, next to the refuge visitors center.

The 100-acre **Bailey Tract,** off Tarpon Bay Road, is a freshwater marsh. Development and mosquito-control efforts have greatly altered much of Sanibel, but here is a slice of the true Old Sanibel. Hiking and biking trails crisscross the land.

The **Perry Tract** is three acres of Gulfside beachfront property, near Gulfside City Park, where native dune vegetation grows without the intrusion of exotic species.

Most refuge land is open to the public. Forty-four percent — 2,800 acres — is federally designated Wilderness Area, restricted to animals only.

Touring the refuge

Before you head into the refuge itself, stop at the "Ding" Darling Center for Education, the new visitors center (no charge). Knowledgeable volunteers can answer nearly any ques-

---

Facing page: Bikers explore Indigo Trail

tion. Maps and brochures are free, as are binocular rent-als. A series of compelling exhibits introduce you to the area, then an orientation video explains the history and purpose of the refuge and shows what you might see. The center also has a first-rate bookstore and gift shop, with many quality field guides, nature books and souvenirs.

Most visitors first drive (or bike) down the newly-paved Wildlife Drive. You'll travel through a mangrove forest while passing large areas of open water. But don't stay in your car. Get out and walk the shore, wander the trails, and be part of this unique natural area. Along the road are hiking trails, boardwalks and an observation tower that let you get a closer look at animals in their natural setting. Volunteer docents in golf carts are often stationed through-out the drive to help visitors and answer questions.

Get an up-close view of the mangrove ecosystem at the Red Mangrove Overlook. The trees themselves are a tangled mess of roots, like a box of extension cords. Look closely at the muddy ground to spot fiddler crabs. A view-ing platform overlooks a bay estuary and many small man-grove islands. Come at low tide to see thousands of wad-ing birds feeding on the mud flats. At higher tide look for birds roosting in the trees on the small islands. Look down into the reddish, shallow water for shells moving on the mud; these are likely to be king's crown conchs. You'll also see hundreds of tiny fish — food for the birds.

A 90-minute tram tour (run by Tarpon Bay Recreation Center, 472-8900) offers an informative, relaxing intro-duction to the Darling Tract. An island naturalist takes

The main entrance to the refuge (above) is on Sanibel-Captiva Road. The Bailey Tract (below) is on Tarpon Bay Road.

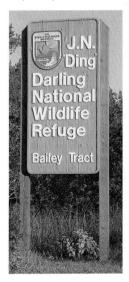

Facing page: Wetlands at the end of Indigo Trail
Overleaf: Mangrove habitat off Wildlife Drive

Above: A visitor at the Red Mangrove Overlook viewing platform looks down at the crabs, conchs and fish underneath

Below: Tram tourers listen to their guide tell stories

Facing page: The Bailey Tract preserves Sanibel's freshwater character

you down Wildlife Drive, calls your attention to interesting habitats and foliage, identifies birds and animals, and sometimes offers his own take on the battle between man and nature. The tram normally makes two trips a day (except on Fridays, call for a schedule). It boards at the visitors center parking lot in summer; the Tarpon Bay lot in winter. The price is $8 for adults, $4 for children ages 12 and under. Reservations are required.

You may spot alligator tracks on Indigo Trail, a two-mile sand path. If you look closely you may see the gator itself lurking in the water. (Note: alligators move very fast and can attack, and kill, small children and even adults. Use common sense and keep your distance.) To explore an inland marsh head over to the freshwater Bailey Tract, which has a network of hiking and biking trails.

Two canoe trails, one off Tarpon Bay and another along the shores of Buck Key off Captiva, take you through red mangrove forests. Manatees live in these areas; watch for one to come up for air.

Building a sanctuary

Established as the Sanibel National Wildlife Refuge in 1945 and administered by the Florida refuge that later became the Everglades National Park, the refuge was re-

named in 1971 for J. Norwood Darling. A nationally syndicated political cartoonist in the 1930s and 1940s, Darling went by "Ding" (a contraction of "Darling") and lived at a seasonal retreat that he built on Captiva. Through his efforts, the sale of state-owned land to developers was delayed, clearing the way for the refuge to be established.

The Bailey Tract was purchased from the Bailey family in 1952, for $50 an acre.

The Wildlife Drive mosquito dike was completed in 1965. It helps control the island's mosquito population by retaining brackish water on the left side of the drive during the summer, preventing salt-marsh mosquitoes from laying their eggs on mud flats.

The refuge was overrun by visitors in the 1980s. Jet skis buzzed in the back bays. Motorcycles and giant tour buses held up traffic on Wildlife Drive. Sightseeing airplanes landed on Tarpon Bay. "At times there was nothing but wall-to-wall cars and buses and mopeds," says former Refuge Director Lou Hinds. "It looked like a city." With the assistance of state and local government officials, Hinds put an end to much of this interference. He shut down large areas of the refuge to human traffic, closed Wildlife Drive on Fridays, created two 450-acre bird retreat zones, limited boat traffic and created no-wake and minimum-speed areas.

Active maintenance

Refuge officers and staff actively maintain the area, removing nonnative plants and trees, restoring native spe-

Heading back from the kayak trails at Tarpon Bay

Go before you go. There are no public facilities in the refuge. The only water fountains and rest rooms are at the visitors center.

Have kids? Ask a volunteer at the refuge information desk for a free nature bingo game to play on Wildlife Drive

Alligators (and the lone Sanibel crocodile) often bask in the sun in the less-saline areas on the left side of Wildlife Drive and along the inland trails

Wildlife Drive was paved in December 2001

Facing page: Rich habitat at the Bailey Tract

Come to the refuge during the peak season and you'll see another strange creature: the wildlife photographer. Station wagons and vans loaded with equipment clog the traffic on Wildlife Drive as their owners fumble with giant tripods and lenses. At times, the birds appear to pose for the camera.

Gated culverts were placed under the Wildlife Drive dike in 1972. They close in the summer to reduce mosquito habitat. They stay open in the winter; the natural tides and mud flats give migrating birds a large foraging area.

Facing page: Volunteer Carolyn Johns awaits visitors on Wildlife Drive

cies, and holding prescribed burns to eliminate thick underbrush and enhance new growth (the burns mimic natural fire cycles). Each year workers chemically treat hundreds of acres of invasive Brazilian pepper and Australian pines, which kill off native plants and destroy habitat.

Hours and fees

The visitors center is open 9 a.m. to 5 p.m. daily from November through April; May through October hours are 9 a.m. to 4 p.m. Wildlife Drive is open from sunrise to sunset every day but Friday. The front gate opens to vehicles 1 hour after sunrise and closes ½ hour before sunset. The Bailey Tract is open sunrise to sunset daily.

You pay $5 per car to drive through the Darling Tract, $1 per hiking or biking group. Golden Age passports, Federal Duck Stamps, Golden Access and Golden Eagle cards and "Ding" Darling annual passes are also accepted. There is no fee for visiting the Bailey Tract.

You pay on an honor system, by placing cash in an envelope at a drop-box at the beginning of Wildlife Drive. One of the most successful honor-fee programs in the U.S. Fish and Wildlife Service (96 percent of visitors pay up), the program is enforced on random days. A refuge officer actually hides in the box and observes violators, who are then stopped at a surprise checkpoint around the bend. Proceeds from the entrance fees support the refuge's environmental education, interpretive and public outreach programs.

Trees & Plants

anibel and Captiva have some of the strangest trees and plants in the United States. One look around you, and you know you're not in Kansas anymore. We have trees that are not trees. We have trees that sweat salt, trees that live up in the air, trees that strangle each other, and trees that walk (or at least look like they do). One of our trees even does impressions — maybe of you.

And though we're thousands of miles from any desert, we have cactus. It climbs in the tree that's not a tree, that's being strangled by the tree that lives up in the air!

In the next few pages we'll tell you the weird truths behind some of the islands' more common trees and plants, and give you tips to find and identify them.

Palms

Have you been to the "Honey I Shrunk the Kids" playground at Walt Disney World? There you're in a fantasy world, among giant pretend stalks of grass, 20 feet high.

Here, it's the real thing. We have grass that's 20, 30 even 40 feet high. You can walk underneath a stalk and stand in its shade. And if a seed falls off, watch out! The size of a football, it can knock you out.

These stalks of grass are palm trees. Or, we should say, palms. (Or, to get really technical about it, "monocot angiosperms.")

Though most people consider them trees, palms are actually more closely related to grasses (as well as pineapples and orchids). Palms have no bark, no branches, no leaves and no wood. Like a blade of grass, a palm is a single fibrous stalk, with green grassy blades shooting out of the top.

Sanibel and Captiva have thousands of palm trees — oops, palms — that give the islands a beautiful tropical character and provide food and shelter for our wildlife.

Cabbage (sabal) palm

A **cabbage palm** *(Sabal palmetto)* looks like an upside-down dust mop. It's the most common palm on the islands; you can't help seeing one. This hearty palm can survive almost

A slice inside a cabbage palm trunk shows the "tree" is made of fiber, not wood
Facing page: A cabbage palm grows in Sanibel's Gulfside Park Preserve

A cabbage palm with boots is so distinctive many people mistake it for a different species. But it's still a cabbage palm, scientifically identical to its bootless brothers.

Growing with a natural curve out of the ground, coconut palms reach up to the sun at the Casa Ybel Resort

anything — when Hurricane Andrew hit Miami, 92 percent of the cabbage palms escaped unharmed. Even most fires don't bother it. The dense fiber trunk protects the moisture deep inside, even if the surface burns. You can find charred cabbage palms alongside Sanibel-Captiva Road (on the refuge side) and on the SCCF hiking trails.

The cabbage palm grows by sprouting large fan-like leaves, called fronds, out of its terminal bud on top. As new fronds sprout, older ones turn brown and droop. In the wild cabbage palms often have many brown fronds hanging from them. They're not unhealthy, just untrimmed. Eventually they fall off, and their bases, called boots, usually fall off, too. (Around streets and buildings, landscapers often cut off the brown fronds for a neater appearance and to control insects.)

Most trunks are smooth. But some are spiky, covered with crisscrossed boots that stay intact.

The palm's berries are a staple of birds and raccoons. Lizards and insects live on the trunk and boots. Cacti and vines, including poison ivy, often grow on the trunk.

Though not really a tree, the cabbage palm is Florida's official "state tree." Also called a sabal palm, its "cabbage" name comes from the fact that its terminal bud looks like a cabbage.

Coconut palm

The icon of the tropics, a **coconut palm** *(Cocos nucifera)* is easy to identify by its clusters of nuts hanging at the base of its fronds. As the palm grows, tall feather-like fronds — each up to 18 feet long — sprout out of the top, eventually falling off and leaving rings around the trunk. The short trunk of a young palm looks like it's wrapped in sheets of burlap.

Each palm has up to a dozen coconut clusters, each with 10 to 20 nuts.

Coconut palms need 40 to 50 inches of rain a year, and temperatures that never drop below freezing. Not originally native to Florida, it was naturalized here sometime before America was discovered by Europeans. Today it is considered a native.

One of the most valuable plants to man, the coconut palm provides food, drink and shelter for people around the world. The sap is a popular drink in some areas, either straight or fermented as wine. Some people chew the roots for their narcotic properties.

Making a Wise Crack
How to open a coconut

Even when it's out of its husk, a coconut is still nearly impossible to open. Many folks have given up trying to crack that tough, brown shell in an elegant manner, and instead just beat the darn thing with a hammer until it shatters. But let us tell you the smart way to do it. You'll not only amaze your friends at parties, you won't get shell fragments in your coconut meat.

You'll need a hammer, a large flat-blade screwdriver and a glass. And you'll need to have a good coconut (by good, we mean one with no cracks that you can feel the milk sloshing around inside). Then it's just a two-step process:

1. Drain the liquid. Hold the coconut firmly on a hard surface. Then, using a hammer, tap a clean screwdriver into one, or two, of the coconut's "eyes," about two inches. Remove the screwdriver and pour out the liquid. (Save it if you want. It's often used in recipes, and some people — not us — drink it straight.)

2. Crack the shell. There's a natural fracture point on the coconut's shell which is easy to find. Place the coconut on a flat surface, and locate a point about a third of the way from the smaller end. Using a hammer, give that spot a light whack. Rotate the coconut slightly, and whack it again, the same distance from the end. Repeat this several times as you rotate it on the surface. Once you see the fracture develop, insert the tip of the screwdriver into it and pry upwards. The coconut should separate in such a way that you can easily get at the white meat.

The whole thing should take three or four minutes, tops.

Facts about coconuts

■ It takes nine to 10 months for coconuts to grow and ripen.

■ The meat in each nut has as much protein as a quarter pound of beef steak.

■ A coconut on a palm looks different from one in a store because it's covered with a husk, just like an ear of corn. (At right, a nut rests inside a husk broken open after a storm.)

■ "Cocos" is Portuguese for monkey. After the husk has been removed, a coconut has three black dots on one end which, as every child can tell you, can look like the face of a monkey.

By the way, don't pick the coconuts growing on the Sanibel Causeway. Those palms have been sprayed with chemicals to prevent lethal yellowing.

Indians in the Everglades still use the fronds to make thatch huts. Fiber from the husk — coir — is used to make rugs, rope, brushes, doormats and mattresses. Besides being eaten, the white meat, called copra, is used to make cooking oil, margarine, soaps and candles. Hawaiians use the swollen base of the trunk to make hula drums. A local couple weaves baskets and hats from the fronds.

In other words, we're cuckoo for coconuts.

A coconut palm grows its trunk out of the side of the nut. The trunk of a native tree will curve in an arc up toward the sun. But many landscapers plant the nuts straight up and down, believing a straight tree is more practical. So when you see a straight coconut palm, you know it came from a nursery.

In fact, most of the coconut palms on the islands have been planted. Years ago, the favorite was the Jamaica Tall, a variety that grows up to a hundred feet high, but a disease called lethal yellowing destroyed most of these in the late 1960s. Today the most popular coconut palm on the islands is the Malayan Dwarf. Resistant to lethal yellowing, it grows up to 30 feet high and starts producing nuts when its trunk is just 3 feet tall.

Above: Young coconuts. **Facing page:** Claudia Mayer weaves a basket from a single coconut frond at Sanibel's BIG Arts crafts fair, using a technique practiced in Florida for over 100 years. The sturdy result lasts for years. **Below:** Royal palm.

Royal palm

Some people say the **Florida royal palm** *(Roystonea elata)* is the most beautiful palm in the world. A smooth gray shaft, often swollen in spots, grows up to 100 feet high. It's topped by a bright-green vertical sheath crowned with fronds, each 10 to 18 feet long.

Often used in area landscaping, royal palms were made popular by Thomas Edison, who imported them from Cuba to line the street in front of his Fort Myers home. Great specimens can be found along Periwinkle Way: in front of the Wood Theatre, around the Bank of America across from Bailey's General Store, and just east of the Old Schoolhouse Theater. You can get a close-up view of the fronds of a royal palm at the Sanibel Public Library, on the west side of the reading porch.

Royal palms are native to Florida, but only grow in the southern third of the state. They can't survive a frost.

Other trees and plants

Gumbo limbo

The **gumbo limbo** *(Bursera simaruba)* has paper-thin, reddish-brown bark that constantly peels; it's said to be doing an impression of a sunburned vacationer. The leaves are clustered at the end of its branches, which spread out in all directions. A gumbo limbo can grow up to 50 feet tall; the trunk can be 3 feet thick. One of the few local trees that sheds its leaves, a gumbo limbo will be barren for one or two months each year.

Unique to coastal southern Florida in the United States, the gumbo limbo also grows in Central America and the West Indies. Its gummy sap can be burned for incense and has been used as a varnish, glue and salve. The tree's soft wood was once used to make carousel horses. "Gumbo limbo" is a corruption of the Spanish name for sap, "gumma elemba."

Look for a large gumbo limbo tree as you walk the Shell Mound Trail at the refuge, close to the boardwalk.

Mangroves

Mangroves are the most important trees on Sanibel and Captiva. The wildlife, and people, depend on them for survival. The bushy, thick trees line the bay shores and inland waters. Mangroves created the islands, keep them from eroding, and provide a remarkable wildlife habitat.

Without mangroves the islands would have no fish, as nearly all fish breed in mangrove estuaries, or feed on the fish that do. There would be no wading birds, as they'd have no fish to eat. In fact, the islands would have nothing at all — no beaches, no resorts, no refuge — because the islands would simply wash away. After Hurricane Donna hit Southwest Florida in 1960, officials noticed that damages were the most severe in areas where mangroves had been trimmed back or cut down. In response, Lee County enacted the state of Florida's first mangrove protection laws. But mangrove trimming is still legal here; if done properly it allows waterfront homeowners to have a view of the water while keeping the trees healthy.

There are three types of mangroves on the islands. Red mangroves grow into the water. Black mangroves are right behind them, at the water's edge. White mangroves and their cousins, buttonwood trees, bring up the rear, grow-

Islanders create new gumbo limbo trees by simply slicing off a branch of a current tree and sticking it in the ground. The new tree grows up to a few feet a year, as the branch becomes a trunk.

Black and white mangroves excrete salt through their leaves. You can often see the crystals.

During the summer, the small flowers of the black mangrove attract swarms of bees. The white honey produced is sold in stores under the name "mangrove honey."

Facing page: The gumbo limbo tree's unique peeling red bark has earned it the nickname "the tourist tree"

The red mangrove is called the "walking tree" because its roots appear to be sitting above the ground, like legs, holding the tree up in the air as it steps out into the water

Two places to see a red mangrove forest up close are the Red Mangrove Overlook at the refuge (above) and along the west side of Dixie Beach Road

Facing page: The staff of life of the islands, a red mangrove seedling washes ashore

ing on the solid ground reds and blacks have created. Salt tolerant but susceptible to cold weather, mangroves only grow in the U.S. on the shores of southern Florida.

The **red mangrove** *(Rhizophora mangle)* grows directly in saltwater. Averaging 20 to 40 feet tall, it grows in dense fringes at the shore. Its "prop" roots appear to sit above the ground, but they actually extend a foot or more into the sand and hold the trees in place.

Thousands of years ago, red mangroves created the thick sandbars that are Sanibel and Captiva. Today, they're keeping the islands from eroding away. An ingenious scheme is at work: The thicket of roots above ground traps leaves, twigs, grasses and silt. This, in turn, prevents waves from eroding the sand. This mass decays to create a denser sand, which becomes the permanent island. Over time the island grows, complete with its own natural anti-erosion system.

A metropolis of wildlife, a red mangrove shoreline is home to everything from shrimp, snails and sea horses to fish, snakes and pelicans. The prop roots give animals a safe, nourishing environment to grow and raise young.

The trees keep the water clean as they filter out waste and pollution. They provide food, too. The fallen leaves are a key part of the diet of crabs and shrimp. Mangroves produce and drop leaves like crazy: each year one acre of trees will produce eight tons of leaves.

Each red mangrove also produces about 300 seeds a year. The seeds sprout into 6-inch seedlings, each with a root and stem, before they fall from the tree. Some drop

like darts, root end first, ready to grow near the parent tree. A lone red mangrove can be surrounded by a forest of others in about 25 years, all from its own seeds. Other seedlings float away, drifting for miles before they land on the shore. Island mangroves are descendants of seedlings that drifted from Africa.

You can spot a red mangrove by its unique roots and the reddish color of the water that surrounds it (the color comes from a pigment in the tree's bark).

You can identify a **black mangrove** *(Avicennia germinans)* by its black trunk and the crystals of salt that it often sweats onto its leaves. But the easiest way is to look down at the ground: It has hundreds of short, pencil-like roots that stick straight up out of the mud. Connected to roots underneath, these "breathing tubes" provide oxygen and help keep the tree stable in the sand.

A **white mangrove** *(Laguncularia racemosa)* looks like a conventional, shrubby tree, with its roots generally in the ground except for a tube or two. Like a black mangrove, it excretes salt through its leaves. Its greenish-white flowers grow on spikes. Its cousin the **buttonwood** *(Conocarpus erectus)* tree grows on higher ground. If a buttonwood falls over (a common occurrence), it can grow a new trunk out of the side of the fallen one. Many old, graceful buttonwoods line the Sanibel River.

A buttonwood tree grows alongside the Sanibel River. It gets its name from its fruit: small berries that resemble vintage shoe buttons.

Periwinkle

The **periwinkle** *(Vinca minor)* flowers associated with Sanibel are not natives — they were planted by the Bailey family long ago. Today they grow wild on the islands, and are still used in landscaping. An easy place to see them is at Gulfside City Park, on the walkway to the beach. Lovely purple and white periwinkles line the path.

The seed stalk of a fishtail palm, common in island landscaping

Facing page: Colorful periwinkles grow wild on the islands. Sanibel's Periwinkle Way was named for these flowers.

Prickly pear cactus

There are no deserts on Sanibel or Captiva, but there are cactus. **Prickly pear cactus** *(Opuntia compressa)* grows along roadsides, in open sandy areas and especially in the sand dunes next to beaches. Easily recognizable by its large pads covered with spines, it's usually about 2 feet tall.

Yellow flowers bloom from spring to fall, followed by a plump, reddish-purple fruit that many islanders use to make homemade jelly. The whole plant is edible. Go-

The Inside Story
Sanibel's unique freshwater wetlands

One reason so many birds seek out Sanibel is because of its abundance of fresh water, a rare feature for an island. There are no springs on Sanibel; the water is simply accumulated rain, settling in the lower areas of the island.

Sanibel has 1,200 acres of freshwater wetlands, pockets of standing water and small swamps in the center of the island. The wetlands have played a vital role in Sanibel's history. They allowed for a thriving wildlife population, which attracted the early conservationists, such as Jay N. "Ding" Darling, who established Sanibel's environmental legacy. Fresh water also allowed a permanent, rural community to form, which provided the manpower to fight Lee County's development plans in the 1960s and 1970s. The U.S. Army Corps of Engineers connected the wetlands with a nine-mile canal (above, today known as the Sanibel River) in the 1950s, so mosquito-larva-eating fish could thrive and spread.

The Sanibel-Captiva Conservation Foundation (SCCF) has worked hard to maintain the wetlands. The group has bought up most of the land, and actively manages it by eliminating exotic plants, restoring ponds and native vegetation and holding prescribed burns. SCCF's 26-acre Pick Preserve, across from the Sanibel School, is used to give island children a hands-on environmental education (at left, third graders).

Today Sanibel is the only island in Florida with natural fresh water. Palm Beach and Miami Beach originally had it, too, before development.

Facing page: A prickly pear cactus blooms in the Pick Preserve along Sanibel-Captiva Road

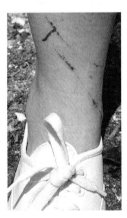

Cuts from saw grass happen when the blades get tangled around your arms or ankles

Saw grass grows in the wetlands throughout Sanibel and Captiva

pher tortoises love to munch on it, spines and all. You'll see good examples of prickly pear cactus on the trails at the Sanibel-Captiva Conservation Foundation and along the walkway to Bowman's Beach.

Seagrape

Before the post office tightened its rules, islanders would use the leaves of the **seagrape** *(Coccoloba uvifera)* tree for postcards. The trees are all over Sanibel and Captiva, and the sturdy, leathery leaves make a decent writing surface.

Though it can grow to 45 feet tall, a typical seagrape is less than half that size. It has a contorted trunk, and oval- or heart-shaped leaves that are up to 11 inches wide. Islanders make jelly from the fruit, which looks like small grapes and grows in late summer and early fall.

Saw grass

Each blade of **saw grass** *(Cladium jamaicensis)* has a serrated, saw-like edge so sharp it can cut into your skin. From a distance saw grass looks deceptively soft and fluffy. Common on the islands, it grows in the wetlands in thick clusters, providing cover and nesting sites for birds. Migrating ducks eat the seeds to restore their energy. Sawgrass is easy to identify in late spring and early summer, when its swaying, reddish-brown seed heads bloom. It's common in the Bailey Tract of the refuge. Not a grass at all (it's a sedge) saw grass has triangle-shaped, instead of round, stems. The leaves sprout tiny flowers.

Sea oats

True amber waves of grain, **sea oats** *(Uniola paniculata)* serve a vital purpose: they bind the sand to the shore, preventing erosion. The most common plant along the dunes, sea oats grow in dense clumps, with stalks up to 7 feet high, topped with flowers and seeds. Strong underground stems anchor them to the sand. Because of their importance to the beaches, sea oats on the islands cannot be picked or removed.

Strangler fig

The **strangler fig** *(Ficus aurea)* tree gets its name from the fact that it wraps itself around a cabbage palm and strangles it to death!

The strangler fig couldn't exist without wildlife. Birds and raccoons love to pick the fruit off a fig tree then sit in a nearby cabbage palm to eat. When a seed leaves the animal's digestive track it starts to fall to the ground, but gets stuck on the palm's ragged trunk and sprouts. The tree grows there, sometimes 15 feet up in the air. Roots head toward the ground, many wrapping themselves around the palm's trunk, preventing it from expanding. Meanwhile, the fig's branches and leaves grow up and above the poor palm, blocking it from the sun.

Within about 20 years the palm dies, and the fig's weave of roots grow together to form a trunk. The fig then becomes a free-standing tree, and the process starts over again.

Strangler figs are easy to identify — just look for a gray-trunked tree literally wrapping itself around an unlucky cabbage palm or tree. Many strangler figs line Periwinkle Way. There's one right on the boardwalk at the Periwinkle Lazy Flamingo restaurant, another in the parking lot in front of Pippin's restaurant, and still others in front of the Episcopal Church.

White stopper

The **white stopper** *(Eugenia sp.)* tree will stop you in your tracks — it smells like a skunk. For some reason, the smell is strongest about 25 feet away. Easy to smell after a rain, this small, stubby tree flowers late in the summer. Actually, the name "stopper" refers to its use as a remedy for diarrhea. Eating its fruit or making tea from its leaves are both said to stop the problem. Look —or sniff — for a white stopper on the trails at SCCF.

Sea oats growing in the dunes near the Tarpon Bay Road Beach

Fruit trees on the islands include banana (above), grapefruit, lemon, lime, mango and orange

Facing page: A strangler fig "hugs" a cabbage palm in front of the Village Shops on Periwinkle Way

Alien Invaders
Nonnative trees and plants destroy paradise

A pop quiz: What destroyed the most wildlife habitat on Sanibel and Captiva in the 20th century? a) Greedy developers. b) Hurricanes. c) Imported trees and plants.

The answer is C. Imported trees and plants have wiped out more natural habitat on the islands than anything else — even those greedy developers! Some exotics, like melaleuca, have already been eliminated. But others, particularly aggressive Australian pine *(Casuarina equisetifolia)* and Brazilian pepper *(Schinus terebinthifolius)* trees, are still here.

An **Australian pine** looks like a tall, fluffy pine tree (it's actually not a pine, and related to poison ivy), and covers much of the islands. Large older trees line Periwinkle Way and Captiva Drive; thousands more grow in out-of-the way areas and along the beaches.

Sanibel has an outstanding overall record of environmental protection, but it's done a lousy job with the Australian pine. By creating dense shade and covering the ground with its needlelike leaves, it has killed off hundreds of acres of palms and other native vegetation, destroying much of the island's ecosystem and wildlife habitat.

In fact, Australian pines have spread over so much of the island, many people are unaware of how Sanibel looks in its natural state. But it is still allowed here. Strict laws protect wildlife and natural vegetation, but there is no law prohibiting this menace that destroys them. Though 150 types of trees and plants naturally grow on the islands, Australian pines kill off an astounding 145. And as native plants disappear, so does wildlife.

The refuge and SCCF remove the trees from their land. But the city of Sanibel still allows it. Some residents love the tree for its beauty, and refuse to see its problems. Unfortunately, this sincere but misinformed group has persuaded officials to keep the trees here.

Floridians introduced Australian pines to the state in the early 1900s. Farmers brought the tree to Sanibel and Captiva in the 1920s to use as a wind buffer, and planted them along Periwinkle Way and Captiva Road. Hotel owners planted more. (Not all islanders were thrilled. "I detest the Australian pines," wrote "Ding" Darling in 1941. "To me they are about as unsuitable as red flannel underwear on a Tahiti native.")

When Hurricane Donna hit Sanibel and Captiva in 1960, the seeds blew throughout the islands. Australian pines began growing everywhere, up to 5 feet a year, up to 150 feet high, taller than anything else.

A few years later the Sanibel Community Association published a booklet that praised the Australian pine, saying it "seems to have provided enough windbreak and humus from the dropping of its needles [sic] to have encouraged the rapid growth of many native things." But actually the tree does just the opposite.

The tree causes other problems, too. Its

Forty feet high in the air, a worker cuts down an Australian pine piece-by-piece, restoring a gopher tortoise habitat

flat, shallow roots are not made for sand, and give way in strong winds. After Hurricane Andrew hit Homestead in 1992, every Australian pine was leveled (most coconut palms lost only some fronds; nearly every cabbage palm survived intact.) At the beach, the tree wipes out the vegetation that keeps the sand in place, causing erosion problems so severe some parts of Sanibel and Captiva have completely washed away. At the west end of Bowman's Beach, exposed roots entangle sea turtles, both the females coming onshore to lay their eggs, and the just-hatched babies on their dash to the sea.

Brazilian pepper was brought to the islands in the 1950s as a landscape plant, and, incredibly, to provide food for wildlife. Many islanders planted it. But like the Australian pine, Brazilian pepper wipes out wildlife by taking over habitat. The berries are toxic to most wildlife, but migrating robins, one of the few animals that eat the berries, spread the problem. The seeds germinate after passing through the bird's digestive tract. Pepper now covers over a third of Sanibel. Also called the Florida holly, it's the dense, bushy green tree, usually about 12 feet high, sometimes with red berries, throughout the islands. As you drive up San-Cap toward Captiva it lines the left side for nearly the entire route.

But this time islanders are fighting back. Now illegal to plant, pepper has to be removed from a property before the owner can get a development permit. The city removes pepper from its land, and a volunteer group, the Pepper Busters, chops it down in other areas.

Florida's Department of Environmental Protection considers both Brazilian pepper and Australian pines "noxious weeds" and prohibits their transport and sale.

Chain of fool. When workers were clearing a small section of Australian pines from Bowman's Beach a few years ago, a misguided "environmentalist" chained himself to a tree and threw away the key, "rescuing" the tree by refusing to let the workers break his chain. Then, as dusk settled and the workers began to pack up for the night, he was attacked by mosquitoes, and demanded that workers break his chain. They left him there.

After Australian pines kill off the sea oats and other plants that hold a beach in place, they fall over into the sea as the beach erodes. With no dunes left, this section of Bowman's Beach, a mile north of the public park, is completely washing away.

Biking

*T*he best way to see Sanibel is on a bike. You'll feel the sun on your back, smell the tropical scents, and hear the calls and cries of the wildlife. Twenty-three miles of paved bike paths run alongside every major road, from one end of the island to the other. Every place you want to go — hotels, beaches, restaurants, theaters, shops, groceries, museums, even the wildlife refuge — is directly on a bike path. The island is almost perfectly flat; even a child can ride for miles with ease.

Where to go biking

You can go anywhere on Sanibel on a bike. But here are seven itineraries to help get you started. They include trips to major sites, attractions and beaches, and into hidden areas. (Circled numbers refer to locations on the bike path map on page 195.)

❶ **Periwinkle Way commercial area.** *(Along the west end of Periwinkle Way. 5.4 miles round trip. Allow 90 minutes for bikes, 30 minutes for scooters, two or three hours for surreys.)* It's Sanibel's main drag, but Periwinkle Way has a rural feel. The bike path often ambles away from the road; the middle section is lined with tropical foliage. The Periwinkle Place Shopping Center is a good resting spot, with a small playground for kids. *Diversion:* Stop at the Periwinkle Park campground and see its collection of exotic animals kept in a small walk-around mini-zoo. Ask permission from the gate attendant before you ride in.

Facing page: Periwinkle Way bike path. **Above:** Family fun in a Sanibel surrey.

On their way to the lighthouse, a father and daughter coast through an S-curve along Periwinkle Way

Checking out the lush landscaping along East Gulf Drive

❷ **Sanibel lighthouse area.** *(Periwinkle Way from Causeway Blvd. (Lindgren Blvd.) to the lighthouse. 2.6 miles round trip. Allow two hours for bikes, one hour for scooters, three hours for surreys; more time if you stay awhile at the beach.)* This trip takes you over canals and to the lighthouse and its beach. There's a drinking fountain at the beach parking lot; restrooms by the lighthouse. *Diversion:* At the four-way stop of Periwinkle and Lindgren, go south on Lindgren to ride through Shell Harbor, a 1960s-era subdivision. Head left (east) down any side street to see vintage concrete-block homes, many with their original rock yards. *Beach access points:* At the lighthouse, as well as at either end of Buttonwood Lane and Seagrape Lane.

❸ **Gulf Drives tour.** *(East Gulf Dr. to Middle Gulf Dr. to Casa Ybel Rd., out to West Gulf Dr. and Rabbit Rd., then return via Sanibel-Captiva Rd. 13.1 miles round trip, plus side trips. Allow at least four hours for bikes, one hour for scooters. Add more time if you stop at the beaches or attractions.)* This calm ride takes you along the front of Sanibel's Gulf-front resorts and past many beach access points. The Rabbit Road leg leaves the roadside and runs along a canal, past two large ponds and over the Sanibel River. Public restrooms are at Gulfside City Park and the Tarpon Bay Road Beach. Attractions along the return route (Sanibel-

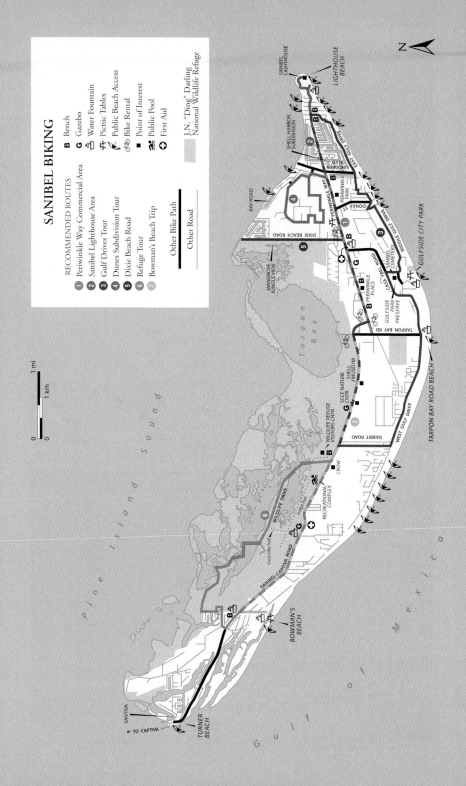

SANIBEL BIKING

RECOMMENDED ROUTES

1. Periwinkle Way Commercial Area
2. Sanibel Lighthouse Area
3. Gulf Drives Tour
4. Dunes Subdivision Tour
5. Dixie Beach Road
6. Refuge Tour
7. Bowman's Beach Trip

B Bench
G Gazebo
Water Fountain
Picnic Tables
Public Beach Access
Bike Rental
Point of Interest
Public Pool
First Aid

J.N. "Ding" Darling
National Wildlife Refuge

Other Bike Path

Other Road

The Rabbit Road trail runs behind homes, along conservation land and across the Sanibel River

Bike-path bridges cross many small streams and canals. Below, a bridge in Gulfside City Park.

Captiva Road) include the Sanibel-Captiva Conservation Foundation (SCCF) and the Bailey-Matthews Shell Museum. *Diversions:* As you ride down Middle Gulf Drive, just after the bike path crosses the road, watch for an unmarked second bike path veering off to the left. This path goes through the wilderness of Gulfside City Park, past the Gulfside Park Preserve hiking trail and the little-known historic Sanibel Cemetery. At Algiers Drive, turn left to go to the beach, or right to leave the park and return to the main bike path on Casa Ybel Road. *Beach access points:* (1) At the first curve on East Gulf Drive; (2) at Beach Road; (3) at the end of Nerita Street; (4) at Donax Street; (5) at the 90-degree turn of Middle Gulf Drive (near Par View Drive); (6) at Gulfside City Park; (7) at the Tarpon Bay Road Beach. There are also seven little-known beach-access points along West Gulf Drive, past Rabbit Road. Watch for a series of paths, most with a tiny parking lot, every tenth of a mile or so. Only island residents can park vehicles here, but it's fine to prop up a bike (not a scooter) along the fences. The beaches are open to the public.

❹ **Dunes subdivision tour.** *(From Periwinkle Way: Bailey Rd. to Sandcastle Rd., and return. 3.2 miles round trip. Allow one hour for bikes, 25 minutes for scooters.)* This tour of Sanibel's main subdivision takes you down

Sandcastle Road, a relatively calm loop street (not a bike path) that surrounds a golf course. Note the odd island architectural features on the newer homes, such as elevated first floors, tin roofs and the abundance of vinyl siding — a must in this salty environment. During the winter, the south side of Sandcastle gets heavy traffic between 3 p.m. and 6 p.m. *Diversion:* As you leave the subdivision turn left at Bailey Road to go down to either San Carlos Bay (there's no real beach here, but it's a nice spot to take a break) or to Bay Road for a shady 1.3-mile trip to a little-known Sanibel bayside subdivision. *Beach access points:* At the end of Bailey Road and at the end of Bay Road.

❺ **Dixie Beach Road.** *(Dixie Beach Rd., off Periwinkle Way. three miles round trip. Allow 45 minutes for bikes, 15 minutes for scooters, not counting diversions.)* This straight line takes you past mangrove habitat to tiny bayfront Peace Park. Most of the trip is on the road itself, the bike path degrades to weeds after a few hundred yards. *Diversions:* Stop at the culvert halfway down for a close-up view of a mangrove jungle. At the bay, turn right to ride past unique waterfront homes, or go left down a small shell road for a taste of rough-hewn island life from an earlier day.

❻ **Refuge tour.** *(From Sanibel-Captiva Rd.: Through the refuge on Wildlife Dr. and return. 8.3 miles round trip. Allow three to four hours for bikes. Scooters cannot enter the refuge. $1 per bike-family group, cash only. Open sunrise to sunset.)* This trip through the J.N. "Ding" Darling National Wildlife Refuge will immerse you in the natural beauty of Sanibel. Take the paved Wildlife Drive for an easy trek, or ride down the packed-sand Indigo and Cross Dike Trails. The air-conditioned visitors center has drinking fountains and restrooms. *Diversions:* East of the refuge entrance is the Shell Museum and SCCF. Just west is the Sanibel Recreational Complex, which has soft drink machines and a community pool.

❼ **Bowman's Beach trip.** *(From Tarpon Bay Rd.: Sanibel-Captiva Rd. to Bowman's Beach Rd. and to beach. 11 miles round trip. Allow one hour for bikes, plus time at the beach and any diversions, up to a full day.)* You'll ride past miles of refuge land on the way to Bowman's, which has two shady picnic areas, restrooms, outdoor showers and a water fountain. *Diversions:* On your way you'll pass the air-conditioned Shell Museum and SCCF, as well as the Recreational Complex. Continue past Bowman's on San-Cap 1.5 miles to reach Santiva and Turner Beach.

Public restrooms along the Sanibel bike paths:

■ Chamber of Commerce, on Causeway Blvd.

■ At the lighthouse

■ The Tahitian Gardens and Periwinkle Place shopping centers, on Periwinkle Way

■ Gulfside City Park, just off Casa Ybel Road

■ Tarpon Bay Road Beach, at Casa Ybel Road and Tarpon Bay Road

■ The visitors center at the J.N. "Ding" Darling National Wildlife Refuge

■ The Sanibel Recreational Complex, next to the pool

■ Bowman's Beach

■ Turner Beach, at the west end of the island

Biking through the wildlife refuge along the Cross Dike Trail

Motor scooters at Billy's Rentals on Sanibel

The Sanibel bike paths were created in the early 1970s. Island kids loved to ride, but a development frenzy was bringing in dangerous traffic. Families formed the Sanibel Bike Path Committee and convinced officials to build bike paths throughout the island.

Riding after a rain is a special treat. Frogs, snakes and tadpoles are out and about; the aroma of the tropical foliage is strong and sweet. Some islanders even ride *during* a shower (the raindrops are warm).

Facing page: Colorful, fringed surreys are a popular way to cruise along Periwinkle Way. Great for couples and families, they make it easy to relax and talk to each other. They're available at Billy's Rentals.

Rental shops

Sanibel bike shops rent bikes and accessories to fit any need. Choose from mountain, hybrid and pro-series bikes, as well as tandems ("bicycles built for two"), surreys, motor scooters, in-line skates and adult trikes. For kids there are bikes, in-line skates and foot scooters; for infants and toddlers there are single or double trail-a-bikes (bicycles with everything but the front wheel that attach to adult bikes), bike trailers and single and double baby joggers.

A four-hour rental costs $5 to $16. Weekly rentals are a better deal, ranging from $25 to $75. A two-seater surrey goes for $32 (for four hours); a four-seater is $48. In-line skate rentals start at $10 for four hours, including protective gear. To rent a motor scooter (starting at $25 for one hour), you'll need a valid driver's license and a credit card, and be at least 18 years old.

Billy's Rentals *(1470 Periwinkle Way, Sanibel; 472-5248)* has 1,300 pieces of equipment, including 450 bikes and 30 one- and two-seat motor scooters. Billy Kirkland's shop is the only place to rent a surrey, and the only spot on Sanibel to rent a scooter (which require a $250 deposit). Rentals of two or more days include free delivery and pickup. Founded in 1975, Billy's is the granddaddy of island renters. It also rents bikes at Casa Ybel and the West Wind Inn.

Tucked in behind Winds beach shop, **Finnimore's Cycle Shop** *(2353 Periwinkle Way, Sanibel; 472-5577)* has 100 bikes to rent, is the only place on Sanibel renting in-line skates, and sells skateboards. Rentals of three days or more include free delivery and pickup.

Bike Route *(2330 Palm Ridge Rd., Sanibel; 472-1955)* targets serious bikers, with a rental fleet of 25 pro-mountain, pro-road and pro-tandem bikes. There's no pickup or delivery, but owner Bill Wallstedt will loan you a car rack for an hour or two, or rent you one for $15 a week. The **Tarpon Bay Recreation Center** *(900 Tarpon Bay Rd., Sanibel; 472-8900)* is the closest bike rental spot to the refuge. The official refuge concession, it has over 30 large-tire, single-speed bikes, all with baskets. Trail-a-bikes and baby seats are available. The only place to rent a bike, motor scooter or pair of skates on Captiva, **Jim's Rentals** *(11534 Andy Rosse Ln., Captiva; 472-1296)* shares space with YOLO Watersports. Scooter rentals require a competency test. Jim's also sells skateboards.

Many hotels and resorts rent bikes, too.

Hiking

*W*ith most of Sanibel preserved in a natural, undeveloped state, there are some great opportunities to get out and explore. There are no tough hikes here; most of the trails are short and the land is perfectly flat. But it does pay to bring insect repellent, sunscreen and a bottle of water, and wear loose clothing, comfortable shoes, a hat and sunglasses. Bring binoculars for a closer look, or a camera to record your adventure. *Boxed numbers below refer to locations on the Sanibel and Captiva maps on pages 6 through 8.*

'Ding' Darling

There is no better way to see the J.N. "Ding" Darling National Wildlife Refuge *(off Sanibel-Captiva Rd., 2½ miles west of Tarpon Bay Rd.; One Wildlife Dr., Sanibel; 472-1100)* than on foot. The long Indigo Trail borders tropical wetlands, the short Shell Mound Trail winds through a hardwood hammock, and a series of trails wanders through the freshwater Bailey Tract. Trails are open to hikers and bicyclists from sunrise to sunset, though Wildlife Drive is closed on Fridays.

Stop in at the visitors center before you set out. The front desk has pamphlets, maps, binoculars for loan, and knowledgeable volunteer docents ready to answer any question.

Indigo Trail 1 runs two miles from the visitors center to the Cross Dike. It starts as a boardwalk winding under mangrove, white-stopper and buttonwood trees. At Wildlife Drive the trail becomes a long sand path through remote land. There's a bench on your left at the halfway point. Indigo Trail is the only part of the Darling Tract open on Fridays. Give yourself two to three hours for the round trip. The boardwalk is a great place to see small, flitting songbirds. The dike has wading birds feeding on both sides, including roseate spoonbills and ibis, and wild coffee trees tucked in under the palms and mangroves. Toward the end of the trail notice the trees thinning out on your right, and the water turning into marshland. Hundreds of wading and shore birds often feed here. The ¼-mile **Cross Dike Trail** 2 rejoins Wildlife Drive, and again has wetlands on both sides. A covered viewing pavilion, built where the refuge's lone crocodile often basks in the sun, has benches and a viewing scope.

For a good hike, start at the visitors center, take Indigo Trail to the Cross Dike Trail, and return via Wildlife Drive. You'll pass a dense mangrove swamp, filled with snails, raccoons and coon oysters. At three miles

Facing page: Indigo Trail. **Above:** The Cross Dike pavilion overlooks a crocodile hangout.

Hiking the refuge.
Top: Bailey Tract trails wander through a freshwater marsh. *Above:* The Shell Mound Trail winds through a hardwood hammock.

Facing page: A guided tour down the Sabal Palm Trail at SCCF

you'll reach the Red Mangrove Overlook, a short boardwalk through the mangroves out to the water. As you walk the final mile, you'll pass many wading bird habitats.

Give yourself 20 minutes to take in the **Shell Mound Trail** 3 , a ⅓-mile boardwalk near the end of Wildlife Drive. Encircling an ancient Calusa Indian shell mound, the boardwalk winds through a rich tropical hardwood hammock, unique in southern Florida. Shady palms and trees arch over you. Look closely to find a small stand of key lime trees, left over from the island's farming past. Interpretive signs identify the different vegetation. Covered with leaves, the ground here is among the driest in the refuge.

Several peaceful but sunny trails roam through the uncrowded **Bailey Tract** 4 *(on Tarpon Bay Rd., between Periwinkle Way and West Gulf Dr.).* This 100-acre freshwater marsh is home to alligators, turtles, snakes, gopher tortoises and many birds. Most trails go in circles; their total length is 1.75 miles. Allow at least an hour here. There's free parking and no admission charge.

SCCF

Don't miss the four miles of shady trails through interior wetlands at the Sanibel-Captiva Conservation Foundation *(3333 Sanibel-Captiva Rd., Sanibel; 472-2329).* We recommend two round-trip trails: the 1.1-mile **East River**

Signs keep you on track
on the SCCF trails

All Sanibel trails have
interpretive maps along
their routes. Above, the
entrance to the Gulfside
Park Preserve.

**You can identify a
gopher tortoise burrow**
by the sand and shells dug
out in front of the hole
(don't step here: tortoises
lay their eggs in this spot)

Trail 5 takes you to a primitive observation tower then runs along the Sanibel River; the 1.3-mile **Sabal Palm Trail 6** winds through a palm forest to Alligator Hole. You can walk the trails on your own or go on a group trip with an SCCF guide at no extra charge. Our choice: take the guided tour — learning about the wild vegetation here is fascinating, and you won't get so nervous when you hear the gators croaking in the distance. Also on the grounds are an education center (with a touch tank for kids), a butterfly house and a native-plant nursery.

Gulfside Park Preserve

Located off a little-known bike path that cuts through Gulfside City Park, the 22-acre **Gulfside Park Preserve 7** is the newest hiking area on the island. It's ¾ of a mile long and takes about 45 minutes to explore.

A footbridge crosses a wetland covered in saw grass and leads you to the circular trail, which winds past a pond, through palm groves, and under strangler figs and seagrape trees. Butterflies and dragonflies dart about, attracted to the lantanas. Interpretive signs identify various plants and trees. There's a picnic table, then a bench halfway around, in beautiful, shady spots. A little farther up you'll walk past booted, un-booted and even scorched cabbage palms.

Sanibel purchased the land in 1996, through a grant from the Florida Communities Trust. Workers cleared it of Brazilian pepper and Australian pines, built bird houses and restored the area to its natural state. The trail is made from crushed coconut husks and cabbage palm boots.

Located between Casa Ybel Road and Middle Gulf Drive, the Preserve is not visible from either. It's near the east end of the bike path that cuts between Middle Gulf and Algiers Drive, by the tiny Sanibel Cemetery. Park at the Gulfside City Park lot on Algiers Drive.

The Preserve is next to **Gulfside City Park 8**, 47 acres of forested wetland and upland habitat. Give yourself another 45 minutes or so here. Most visitors only see the few acres of the park next to the parking lot and beach. But hidden back along the bike path is a natural paradise. Gopher-tortoise burrows are easy to see.

Primitive trails wander off the path. One follows a stream; another is the old Middle Gulf Drive, once the main route for mule-pulled wagons to bring vacationers to Casa Ybel.

Silver Key

Another new addition to Sanibel's public land, this 64-acre Gulfside island is located within Clam Bayou and Old Blind Pass at the west end of Sanibel. A 1,200-foot path [9] leads to a small bayou dock, but the real hike is getting here — it's a good mile up Bowman's Beach from the parking lot. Give yourself a couple hours.

This island-within-an-island "is managed to preserve and enhance wildlife and vegetation habitats and to provide environmental education programs to highlight this unique resource and the importance of its protection."

At least that's what the sign says. The truth is, Silver Key is infested with Australian pines, which have killed off nearly every palm and other native plant that once grew here. Most of the wildlife is gone too. Silver Key and Clam Bayou were once a true tropical haven, and hold promise to again be a unique Gulfside estuary. But a serious restoration effort has to happen first.

To get to Silver Key from Bowman's Beach, walk west down the beach, through some fallen pines and across the Clam Bayou inlet (if it's open come at low tide). From Turner Beach walk east, eventually across a wide sandbar that blocks the other inlet. A large interpretive sign, set back in the trees, marks the start of the trail.

Wear socks when you hike through the Gulfside Park Preserve. Saw grass here can cut your skin.

Once you cross the wooden bridge on the path to Bowman's Beach, there's a little-known old shady path that veers right (west). It goes for nearly a mile. Taking this route to Silver Key is a chance to get completely away from civilization.

The Clam Bayou and Old Blind Pass inlets have been blocked by sandbars nearly every day since 1988, leading to the death of most sea life and many mangroves. Storms break open the small inlet occasionally.

The Silver Key trail leads to a Clam Bayou overlook

Boating

*P*ine Island Sound — the bay waters to the east and north of Sanibel and Captiva — is a picturesque haven. Osprey nests top channel markers and abandoned fishing shacks sit far out from shore. There are more dolphins here than any other spot in the Western Hemisphere. And a handful of remote out islands make fascinating rest stops or destinations.

For boating adventures on a smaller scale, consider Sanibel's Tarpon Bay, where the marked Commodore Creek Water Trail winds through a mangrove forest. Canoe and Kayak magazine has named it one of the top 10 places to paddle in the U.S. The bay itself, a breeding ground for manatees, is dotted with tiny islands. For a back-to-nature freshwater trip take a canoe or kayak down the narrow Sanibel River, which meanders through the island's central wetlands.

Watersports are well represented here, too. Two Captiva outfitters let you get up close and personal with the sea on a Yamaha Waverunner, the modern, sit-down equivalent of a Jet-Ski. They'll also take you parasailing, a surprisingly effortless — but thrilling — way to soar high over the tropical landscape.

Pine Island Sound

Group tours

Captiva Cruises *(11401 Andy Rosse Ln., Captiva; 472-5300)* has popular one- to 1½ hour wildlife, natural history and sunset cruises off Captiva Island, as well as destination cruises to Cabbage Key, Useppa Island, Cayo Costa and Boca Grande. Prices are $17.50 to $35 for adults, $10 to $17.50 for children. Reservations are required. Lunch and dinner cruises, party boats, nature tours and sailing excursions are also available. On a smaller scale, the **Sanibel Marina** *(634 N. Yachtsman Dr., Sanibel; 472-2723)* hosts the intimate Stars and Stripes excursion boat.

Charters

Sanibel and Captiva have a wealth of experienced charter guides, all licensed by the U.S. Coast Guard. Each trip offers a different experience, but they do share one thing in common: you're nearly guaranteed to see dolphins.

Sailing from the 'Tween Waters Marina, **Mike Fuery's Tours** *(466-3649)* offer 3-hour nature tours and party boats; **Capt. Jim's Charters** *(472-1779)* have 2- to 4-hour nature tours and lunch cruises, with longer trips available; and **Capt. Randy's Fishy Business**

Facing page: Renting a boat at 'Tween Waters Marina, Captiva

Boating tips

■ **Red right return:**
When you return to
shore, keep the red
buoys on your right,
the green on your left.
This keeps you in the
deep water channel.

■ **Read your chart!** Pine
Island Sound waters
are incredibly shallow.

■ **Wear polarized
sunglasses.** They cut
the glare and let you
see into the water.

■ **Store water-sensitive
items in waterproof
bags and baggies.**
Your boat's interior is
likely to get splashed
with saltwater.

■ **As you cruise the
water,** look behind you.
Dolphins love to surf a
wake. We've even seen
them trailing kayaks.

■ **To keep a pod of
dolphins** from leaving
when you approach,
don't head straight at
them. Instead, come up
beside them, going the
same direction.

■ **Rental rates do not
include** tax, gas or oil.
These charges will add
$50 to $75 to a day on
the water.

Charters *(472-2628)* provide 2- to 8-hour
dolphin watches and snorkeling trips, plus
breakfast and lunch trips to North Captiva,
Cabbage Key, Useppa and Boca Grande.
Randy promotes that he's "kid friendly."

Adventure Sailing Charters *(472-7532)*
is based at the South Seas Resort. The 30-
foot Adventure has room for six passengers
of all ages; larger groups can be accommo-
dated with a second boat. The captain will
urge you to take the controls. Bring food and
drinks; the boat has a refrigerator and head. Every charter
is private, so the boat sails wherever you want to go. The
rate is $95 per hour with a two-hour minimum. Full-day
sails (7 hours or more) are discounted 20 percent.

Sanibel charters (most operate from the Sanibel Ma-
rina) include **Adventures in Paradise** *(472-8443)* party
boats, tours, snorkeling trips and lunch and dinner cruises;
Capt. Brian Holaway's **Around the Sound Tours** *(849-
8687)* through tidal creeks, bayous, keys and islands;
Sanibel Island Adventures' *(826-7566)* 47-ft. Adventure
Cat catamaran, which it charters for sailing throughout
Southwest Florida (catering and crew optional); Capt. John
Gaffney's **Sanibel Island Cruise Line** *(472-5799)*, which
offers sightseeing, snorkeling and shelling trips with coolers, ice, beach chairs, beach umbrellas, towels, snorkeling
gear, shell nets and shell bags provided free of charge; and
Viking Voyages' *(472-6946)* 33-foot pontoon boat, which
has a private upper deck with two chaise lounges (even a
gas-fired grill is available) for cruising, shelling, nature
photography and light fishing (note: no credit cards are
accepted). Sanibel's Castaways Marina is the home base
of **Capt. Joe's Charters** *(472-8658)*, which offers 2-hour
to all-day nature tours and lunch or dinner cruises.

Boat rentals

Renting a power boat can be the highlight of your vacation. Anyone can do it: the boat hand will show you how
to operate the boat, understand navigation buoys and
read a chart (water map). Boats are typically 16 to 21
feet long, with a center console and outboard motor. Most
have a cooler, so bring drinks and snacks along. Look for
a boat with a Bimini top, which shades the seating area.

Captiva generally has the pricier rentals, but you start
off much closer to the out islands. The newest, nicest
rental boats are at **Sweetwater Boat Rentals** *(at the 'Tween
Waters Inn, 15951 Captiva Dr.; 472-6336)*, but the rates
reflect it: half-day rentals are $160; full-day rentals are

The Out Islands

Part of Captiva until a 1920s hurricane, **North Captiva** *(also called Upper Captiva, north of Captiva across Redfish Pass)* today is a reclusive haven for an eclectic group of home-owners, some of whom live here year-round. Worth a stop is Barnacle Phil's, a small restaurant in Safety Harbor. Try the black-bean soup.

One of Florida's largest uninhabited islands*, nearly all of **Cayo Costa** *(also called La Costa, 3 statute miles north of Captiva)* is a Florida state park — the least visited in the Sunshine State. It has nine miles of gorgeous beach, with clear, calm water. Millions of shells lay in the sand. We find sand dollars by the handfuls every time we're up here. Cayo Costa has a basic dock back in a bayside lagoon where you can take the state's free dockside tram (pulled by an old farm tractor) across to the Gulf. Many rental boaters simply pull up on the bay side of the island's sandy southern tip and walk around to the beach.

Up on higher ground you'll find a variety of wildlife, including feral pigs brought over by 16th-century Spanish explorers. An old trail leads to a pioneer cemetery — once the island was connected to a booming phosphate industry on Boca Grande.

The state park area includes picnic tables, grills and two picnic pavilions, as well as a primitive campground with 12 tiny, rustic cabins (for reservations call 964-0375). The park has bathrooms, showers and drinking water, but no electricity, food or other supplies. Bring what you need, including bug spray. Pets, intoxicants and firearms are not allowed on the island. At Cayo Costa's north tip, Boca Grande Pass is world-famous for tarpon. Anglers also catch flounder, snook, redfish, trout, snapper and sheepshead here.

By the way, "Cayo Costa" is Spanish for "island by the coast." Pretty imaginative, huh?

The 85-acre **Cabbage Key** *(east of Cayo Costa, adjacent to Intercoastal Waterway marker #60; follow the marked channel to the small marina)* is home to the Cabbage Key Restaurant and Inn *(breakfast, lunch and dinner daily; 283-2278)*. Built in 1938 by playwright Mary Roberts Rinehart atop a Calusa Indian shell mound 38 feet above sea level, the buildings are still in their original state, as the weathered hardwood floors attest. Jimmy Buffett wrote "Cheeseburger in Paradise" after eating here. Previous diners have pasted thousands of dollar bills to the walls and ceilings of the restaurant; each has a personal message scrawled across the face (look for the ones from Julia Roberts and "Simpsons" creator Matt Groening). The tradition began in 1941, when a fisherman signed and taped his last dollar to the wall to assure that he'd still have money for beer when he came back. Money that falls off the wall is donated to local children's charities and marine research. After lunch, climb the water tower to get a terrific view of this tropical nirvana, then hike through the middle on a winding nature trail. Cabbage Key is named for its many cabbage palms, but it's also dense with bougainvillea and citrus, mango and royal poinciana trees.

Captiva Cruises *(472-5200)* brings a lunch excursion to Cabbage Key daily. If you're taking your own boat, get here before the Captiva Cruises boat to avoid a wait. Jensen's Marina *(472-5800)* offers a round-trip water taxi for $130 for up to 6 people.

Though the entire 100-acre **Useppa Island** *(east of Cabbage Key)* is a private resort, day visitors are allowed to stop in through a special Captiva Cruises lunch tour. The restaurant is in the former vacation home of Barron Collier, who bought the island in 1912. A historical museum has displays on the Calusa and the U.S. Bay of Pigs invasion of Cuba — the CIA trained Cuban nationals here in 1961. When you dock look for multi-million-dollar yachts in the marina; sometimes there is a 50-year-old wooden Trumpy here.

*Well, almost. One eccentric elderly woman lives here.

You can launch a canoe, kayak or motorized boats under 14 feet off Wildlife Drive. Site No. 1, near the refuge entrance, is a non-motorized zone (boats with motors can launch, but must pole or paddle out). Launch site No. 2, just past the observation tower, is in a motorized zone. The water at both spots is clear and teeming with fish. Prime manatee habitats, all refuge waters are zoned Slow Speed/ Minimum Wake.

Former Secretary of the Interior Bruce Babbitt has called Tarpon Bay "a shining example of how a concession can be done right." Below, kayakers Vince Burkhead and Lisa Smith travel the Commodore Creek mangrove trail.

$275. **Jensen's Twin Palms Marina** *(15107 Captiva Dr.; 472-5800)* offers a variety of rentals, from a 14-foot skiff to a 24-foot pontoon boat that holds up to 12 people (call for rates). **Seawave Boat Rental** *(at the South Seas Resort Bayside Marina, 5400 Plantation Rd.; 472-1744)* offers boats for two to 10 people.

On Sanibel the majority of rental business is handled by the **Sanibel Marina** *(on the east end at 634 N. Yachtsman Dr.; 472-2531)*. It has nice power boats for $100–125 for a half day, $175–200 for a full day. Also doing a good business is the tiny **Castaways Marina** *(at the west end of the island at 6460 Sanibel-Captiva Rd.; 472-1112 or 800-375-0152)*.

Inland waters

The official concession of the J.N. "Ding" Darling National Wildlife Refuge, **Tarpon Bay Recreation Center** *(900 Tarpon Bay Rd., Sanibel; 472-8900)* rents kayaks and canoes for exploring the Commodore Creek mangrove trail and Tarpon Bay (the entire area is refuge land). Rates are $20 for two hours; $5 each additional hour. The center also offers guided tours: you can paddle with a naturalist through the water trail (with a group, or with just you and a guide in the same kayak) or take a Sunset Paddle out to rookery islands to watch hundreds of tropical birds come in for the evening. (Kayaks are the boat of choice here. They're more stable than canoes, as they sit lower in the water.)

Captiva Kayak & Wildside Adventures *(11401 Andy Rosse Ln., Captiva; 395-2925 or 877-EZ-KAYAK)* rents kayaks and canoes by the hour, half-day, day and week. It offers guided kayak and canoe trips at sunrise, sunset, even at night (try it during a full moon). **Canoe Adventures** *(Sanibel; 472-5218)* offers canoe tours with former Sanibel mayor Mark "Bird" Westall. Head down the Sanibel River, through the refuge or to Buck Key, an undeveloped bay island a stone's throw from Captiva. The founder of the International Osprey Foundation, Bird is an expert on local flora and fauna.

Sailing, sailing. A handful of Captiva charter captains (above) will take you sailing in the Gulf or bay. Rent your own boat at Captiva's Wildside Adventures (395-2925). Rates for a Windrider trimaran are $35 hourly, $100 for four hours, $175 for 8 hours. The larger Gulf-front resorts rent boats, too. Learn to sail at the Offshore Sailing School (below, 454-1700 or 800-221-4326). Based at the South Seas Resort, it offers one- to seven-day courses for any skill level.

Watersports

Take a Waverunner cruising in the Gulf at **YOLO Watersports** *(11534 Andy Rosse Ln., Captiva, 472-YOLO)* for $60 for a half-hour; an hour is $85. Guided one-hour tours are $120. Prices are per Waverunner, with no charge for additional riders. Renters have to be at least 18 years old, but drivers need to be only 16. Passengers must be at least 44 inches tall. Twenty-five thousand people have parasailed with YOLO ("You Only Live Once") since the mid-1980s, including actress Jane Seymour and her twin boys. The smooth flights feature a dry takeoff and landing from the boat. Per-person prices start at $45; you can fly single, double or triple. Non-flyers can ride in the boat ($10) if there's space. Trips start at 9 a.m.; each is on the water an hour to 90 minutes.

Holiday Water Sports *(in the South Seas Resort, Captiva; 472-2938 or 472-5111, ext. 3433)* is open to the public. To get past the South Seas gate simply tell the guard your plans. You'll get a pass to proceed to the north end of the resort at the T-Dock. Waverunner rates start at $35 for a half-hour on a single-rider machine; three-person Waverunners are available. Guided "Waverunner Safaris" start at $100 per person. Parasailing prices start at $60 for a single rider on a 600-foot line (teenagers get a break on Wednesdays, when their rate is cut to $45). Observers can ride in the boat ($10), but may be bumped for flyers. Holiday Water Sports also rents hydrobikes, sailboats, windsurfers, kayaks and canoes. Other organized activities include water skiing, banana boat rides and sailboat and windsurfing lessons.

Adventures at Sea

We had a terrific time on Holiday Water Sports' Waverunner Safari, a two-hour trip around the out islands. Afterward we jotted down our experience:

We meet at the beach. Our group has seven people, three riding on their own Waverunners and four people doubling up on two machines. As we put on our life jackets, our guide shows us how to operate the machine, tells us what to look for on our trip, and teaches us some basic hand signals so we can communicate with each other while we travel.

Like a motorcycle gang, we climb on our "bikes" and head out. Once we're out in the Gulf our guide gives us 10 minutes to zip around to get used to the machines. We learn to speed up to turn (a Waverunner steers by propelling water through its nozzle jets) and to crouch when the water gets rough.

Then we're off. As we cruise down the Gulf side of North Captiva we spot a pod of dolphins. Our guide slows us down, and we each roam through the herd. There are about six or seven dolphins in the group, including some youngsters that jump all the way out of the water. We are so close — sometimes just a few feet away — we can hear the sound of their blowholes when they surface. It is breathtaking.

We cruise around the Gulf, then head over to Pine Island Sound. We stop for a rest (above) and our guide tells us the history of the area, complete with some wild pirate tales. We watch a manatee surface, then ride over to an abandoned fish camp — an old shack out in the water, on stilts. On the way back to South Seas the water is so smooth our guide leads us through a series of tight "S" turns. We carve up the sea with glee.

Parasailing (right) is another unforgettable experience. We've gone with YOLO quite a few times. You meet your guides at the beach (the guides look like they live in the sun, with bleached hair and bronzed skin), then wade out to the boat and climb in.

The driver zooms out to deeper water, and then comes those exhilarating words: "Your turn!" You're fitted with a harness around your waist and thighs; it feels like a swing. A large parachute is secured to its top, the front is clipped on to a long rope (wound, for the moment, down inside the boat), and you waddle onto the large carpeted launch pad. The driver slowly accelerates, the parachute fills with air, and instantly you lift off.

As the line spools out you soar higher and higher, eventually 500 feet or more above the water. It's quiet up there, and calm.

And the view is spectacular — not just of the islands, but into the water. We saw some huge rays swimming under the surface. Eventually the boat reels you back in, dipping you into the water once if you ask for it.

Note: Young kids can go parasailing, too. They ride with a parent, strapped in front.

Fishing

*B*ait casting, spin casting, fly fishing — the islands offer nearly everything an angler loves. "Sanibel and Captiva are phenomenal places to fish," says Sanibel resident Steve Alexander. "The water is pristine and virtually unpolluted."

"I don't care what I catch," adds Richard Monaghan, visiting the islands from Tulsa, Okla. "I go out in Pine Island Sound and get a half dozen species of fish, including snook, redfish and trout." Light tackle is usually all you need, even for grouper, snapper and tarpon. True to their environmental heritage, islanders fish their waters with care. Catch-and-release is the norm, as anglers respect the fish and the future. There are more fish, too, thanks to a recent net ban.

Where to fish

Pine Island Sound has been a haven for fishermen for ages. It has more species of fish than any other spot in Florida. Captiva's Redfish Pass is the top shore-fishing spot on the islands. Just to the north, Boca Grande Pass is the tarpon fishing capital of the world.

You'll catch your limit out at sea, too. Snapper and grouper hang out in droves just three miles off the coast of Captiva, on a series of rocky underwater ledges.

Many people fish right on the beach. Even tarpon swim just a few feet offshore. Other favorite spots include the Blind Pass bridge between Sanibel and Captiva (when the pass is open) and the Sanibel lighthouse fishing pier (open 24 hours a day). The Sanibel Causeway is a popular casting spot.

Tackle, bait and licenses

Light tackle, using a line weight of 6 to 20 pounds, is fine for almost anything but large tarpon and offshore fishing. Many anglers catch their own bait, using a cast net on the grassy flats or off the bridges or causeway. You'll need a Florida fishing license for any fishing unless you're with a licensed charter captain; the charter price will also usually in-

Left: Cleaning a catch on the lighthouse pier.
Right: Beach fishing at Gulfside City Park.

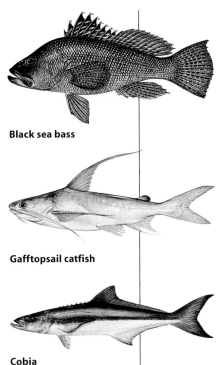

Black sea bass

Gafftopsail catfish

Cobia

The skull of the gafftopsail catfish is known as the crucifix shell. It looks like Christ on the Cross. The fish is said to be the one Christ used to feed the masses.

It's possible to mistake the cobia for a small shark or a snook. The front of its dorsal fin can stick out of the water when the fish cruises at the surface, like that of a shark. Like a snook, a cobia has a stripe straight from its snout through the eye to the tail; unlike the snook, the dark stripe is surrounded with two paler stripes.

clude all your tackle and bait. Just bring a good pair of sunglasses.

The Bait Box *(1041 Periwinkle Way, Sanibel, 472-1618)* is a great bet for tackle and bait. The small shop is run by legendary guide Ralph Woodring. **Bailey's General Store** *(2477 Periwinkle Way, Sanibel, 472-1516)* has a good fishing department, too, with a knowledgeable staff.

What you'll catch

What bites your hook depends on the season. Black drum and sheepshead are common in the winter. Sea bass, cobia, redfish and tarpon are plentiful in early summer. Spring and fall bring mackerel.

Bass

Also called the striped bass, the **black sea bass** *(Roccus saxatilis)* has a jutting lower jaw and a heavy belly. It's dark brown or black, with darker stripes running the length of the body. The dorsal fin has a pattern of white-on-black stripes. *Average size:* 1 to 2 pounds. *Maximum size:* 5 pounds. *Season:* Warm weather. *Where found:* Structures offshore. *Bait:* Mullet, crabs, spoons, jigs. *Edibility:* Excellent. Fine for baking.

Catfish

Also called a sailcat, the **gafftopsail catfish** *(Bagre marinus)* has a dorsal spine that looks likes a sail when extended. A slimy mucous film covers the skin. The dark steel-blue top fades to white on the belly. Long barbels ("whiskers") extend under the mouth. The pectoral and dorsal fins are barbed; the poison they emit can cause a painful wound. When hungry the sailcat can be aggressive and drive other fish out of its territory. *Average size:* 1 to 5 pounds. *Maximum size:* 8 pounds. *Season:* Year-round. *Where found:* Saltwater, but will enter brackish waters. Often caught along docks, bridges and the fishing pier. *Bait:* Minnows or worms. *Edibility:* Good. Pink, sweet meat. Nip the spines before cleaning to avoid injury. Skin before cooking.

Cobia

The bullet-shaped **cobia** *(Rachycentron canadus)* has a broad head and snout, with a protruding lower jaw. The sides have brown and white stripes. The back is dark brown, sometimes nearly black, while the belly is cream-colored. Fins are dark brown to black. Colors are most intense in younger fish. The spines in the front dorsal fin are separated. One of the largest, strongest fish in the flats. *Average size:* 15 to 20 pounds. *Maximum size:* 70 pounds. *Season:* Most of year. Near the islands whenever the surrounding waters are at least 68 degrees. *Where found:* Gulf and bay, submerged structures such as channel buoys and bridge supports, floating objects including floating grasses. Cobia travel in schools of a dozen or more fish, in a variety of sizes. Sometimes cobia swim alongside schools of rays. *Bait:* Live or cut bait; lures. *Edibility:* Smaller fish (under 20 pounds) are excellent. Meat is white with fine grain, similar to grouper. Note: Always skin a cobia before cooking. A second layer of small scales is embedded in a layer of skin beneath the surface scales.

MERISTAR

You don't need a boat to catch your dinner

Drum

The short, solid **black drum** *(Pogonias cromis)* has a body similar to a sheepshead, and a small "beard" of barbels on its lower jaw. Scales have a dark, metallic sheen. Older fish are blackish; young are more silvery and have vertical bars. *Average size:* 3 to 5 pounds. *Maximum size:* 50 pounds or more. *Season:* December through March. *Where found:* Bay, rivers, offshore. *Bait:* Crabs, cut bait. *Edibility:* Coarse, almost tasteless meat, best used for chowder. Worms and parasites are common, though harmless.

Drum get their name from the "bub-bub-bub" sound they make underwater as they contract their muscles over their swim bladders, organs which give them buoyancy

The **redfish** *(Sciaenops ocellatus),* or red drum, has a copper, reddish or gray body. The head is blunt. Most redfish have a large black spot on either side near the tail fin. Some have no spot; others many. The elongated body resembles that of the black drum, but redfish lack chin barbels. Younger reds, known as "red rats," feed on crabs and other invertebrates around oyster bars and mangroves. Larger bull reds are typically found in the passes and offshore. Redfish are also

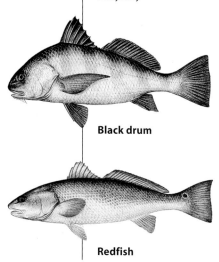

Black drum

Redfish

Cooking Your Catch
Island seafood recipes

Crispy Snapper with Confetti Veggies

By Kevin Pierce

1½ tablespoons olive oil
1   clove garlic, minced (1 teaspoon)
1   red bell pepper, cored, seeded and minced
1   yellow bell pepper, cored, seeded and minced
1   small zucchini, cut in half lengthwise, seeded and minced
1   small yellow squash, cut in half lengthwise, seeded and minced
¼   teaspoon saffron threads, soaked in 3 tablespoons hot water
1   teaspoon chopped fresh thyme (or 1 teaspoon dried)
    Salt, freshly ground black pepper and cayenne pepper
1½ pounds snapper fillets, skins left on
½   cup (approximately) flour

Preheat oven to 400°. In a nonstick frying pan, heat ½ tablespoon oil. Add the garlic, peppers, zucchini and yellow squash; cook on high heat for 1 minute. Mix in the saffron, thyme, salt, pepper and cayenne. Reduce heat to medium; cook the veggies for 3 to 4 minutes until tender, but not soft. Adjust the seasoning with salt, pepper and cayenne to taste. The mixture should be very flavorful.

Cut fish into 4 pieces and season with salt and pepper. Dredge in flour, shaking off excess. Heat 1 tablespoon oil in an oven-safe, nonstick frying pan. Add the fish, skin side down, and cook over medium heat until the skin is very crisp, about 3 to 4 minutes. Flip the fish and place the pan in the oven. Bake for 10 to 15 minutes until cooked. (When done, it will flake easily when pressed.) Place the fish, skin side up, on serving dishes. Spoon the veggie mixture around it and serve at once. Serves 4.

Fish in Coconut Milk

By Jean Baer

4   fillets of fish
    Freshly ground black pepper
2   cups milk
3   large tomatoes, sliced thin
    Salt
1   cup fresh grated coconut
2   large onions, cut in rings

Sprinkle fish with salt and pepper on both sides. Arrange in a single layer in a buttered casserole. Refrigerate until ready to bake. Rinse coconut under cold water, combine in a saucepan with milk. Bring to a boil, remove from heat and let stand at room temperature for 30 minutes, then run the mixture in a blender or strain pressing out all liquid. Spread the onions over the fish, cover with tomatoes. Pour the coconut milk over all. Bake in a preheated oven, 375° for 45 minutes. Serve from casserole. *Do not use "sugar added" coconut.*

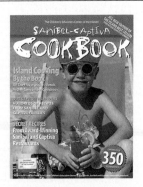

The recipes above are from the Sanibel-Captiva Cookbook, a perennial island bestseller full of recipes from local families and restaurants. It's sold at stores throughout the islands. Proceeds benefit the Children's Education Center of the Islands, Sanibel and Captiva's nonprofit preschool.

known as "red bass" or "channel bass." But they're drum, not bass. *Average size:* 3 to 8 pounds. *Maximum size:* 30 pounds. *Season:* Best mid-April through July. Year-round for smaller fish. Larger bull reds common in spring and fall. *Where found:* Inshore, bay, grass flats, the pier at the lighthouse, along Wildlife Drive, at the mouth of the Caloosahatchee River. *Bait:* Shrimp, lures, minnows, crabs. *Edibility:* Small fish up to 12 pounds are excellent. Large reds have coarse meat and only fair flavor. Most fishermen fillet their catch, but redfish is also good stuffed and baked whole.

Grouper

The **black grouper** (*Mycteroperca bonaci*) has dark blotches on the sides of its body. Many anglers mistake it for the gag. Like all grouper, it has a large head and a broad, down-slanted mouth. *Average size:* 3 to 15 pounds. *Maximum size:* 50 pounds or more. *Season:* Year-round. *Where found:* Offshore for large fish; inshore for small. *Bait:* Live or cut bait, jigs, chum. *Edibility:* Excellent.

Formerly known as the jewfish, the **goliath grouper** (*Epinephelus itajara*) can weigh up to 700 pounds, and be four to five feet long. Young fish have bright spots and mottling on their sides. These details fade on larger, older fish, which have a patchy, blackish-brown color and small eyes. Unlike other grouper, it has a curved tail. The goliath grouper is protected from harvest in Florida waters. If you catch one you must release it. *Average size:* 20 to 100 pounds. *Maximum size:* 700 pounds or more. *Season:* Year-round. *Where found:* Nearshore around docks, in deep holes and on ledges. *Bait:* Spanish mackerel, bonito, jack cravalle.

The **red grouper** (*Epinephelus morio*) is the main man, the grouper served in restaurants. It's reddish brown and splotchy over the entire body (including the fins). The mouth has a brilliant scarlet color when open. *Average size:* 2 to 5 pounds. *Maximum size:* 20 pounds. *Season:* Year-round. *Where found:* Offshore. *Bait:* Live or cut bait; jigs. *Edibility:* Excellent. Firm, white, tasty meat. Widely used for chowder when not steaked. It should always be skinned, as the skin is tough and strong-flavored. Note: Some red grouper have worms. Watch for them when cleaning.

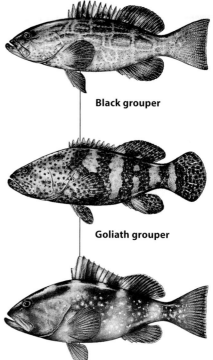

Black grouper

Goliath grouper

Red grouper

The chameleon-like black grouper can lighten or darken its color to suit its environment. You can see this change when you put the fish in a white-sided live well.

'Great snapper!' Florida restaurants serve more red grouper (under a variety of names) than any other fish. The author once worked at a restaurant in Tallahassee. When the kitchen would run out of snapper, it would secretly substitute grouper. Customers often swore it was the best-tasting *snapper* they'd ever had!

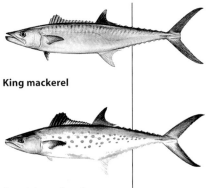

King mackerel

Spanish mackerel

Permanent residents.
These fish are in the
island waters year-round:

- Gafftopsail catfish
- Grouper
- Mullet
- Pompano
- Seatrout
- Snapper
- Snook

Fishing boats at the
Sanibel Marina

Mackerel

The long, streamlined **king mackerel** *(Scomberomorus cavalla),* or kingfish, flashes an overall green and silver iridescence. It has tarnished silver sides with no spots, except when very young, when it has yellowish spots like that of the Spanish mackerel. The tail fin is deeply forked. Far larger than either the related Spanish mackerel or cero, it also should not be confused with the yellow dotted and spotted cero. *Average size:* 3 to 8 pounds. *Maximum size:* 40 pounds. *Season:* Spring and fall migrators. *Where found:* Offshore, inshore near submerged structures such as channel buoys and bridge supports. *Bait:* Live bait, jigs, spoons, chum. *Edibility:* Excellent filleted or cut as steaks. Oily. Spoils quickly. Keep well-iced from the moment caught.

The handsome **Spanish mackerel** *(Scomberomorus maculatus)* has silver sides with brilliant large bronze dots and a silver belly. The back is iridescent greenish-blue, the mouth fairly large with small needle-sharp teeth. *Average size:* 2 to 4 pounds. *Maximum size:* 20 pounds. *Season:* Spring and fall, some during the summer. *Where found:* Gulf and bay, the pier at the lighthouse. *Bait:* Live bait, small jigs, spoons. *Edibility:* Good table fish. Easy to fillet, but, like king mackerel, spoils quickly. Should be iced as soon as caught on all-day trips in high temperatures.

Mullet

Up to 18 inches long, the slender **black (or striped) mullet** *(Mugil cephalus)* is common around Sanibel and Captiva. Mullet is the most widely used bait in Florida, as all game fish seek it. Juveniles, often used as bait for tarpon, stay in Pine Island Sound and other area estuaries until at least six months old. Mullet have rounded heads with small mouths and large scales with dark centers. *Average size:* Fingerlings to 5 pounds. *Maximum size:* 5 pounds. *Season:* Year-round. *Where found:* Inshore. *Bait:* Being vegetarians, mullet are hard to catch by hook and line. Cast nets and gill nets work best. Some anglers use cornmeal balls. *Edibility:* Excellent smoked; good broiled or fried.

Pompano

A member of the Jack family, the silvery **Florida pompano** *(Trachinotus carolinus)* has a gray-blue back and dark upper fins. The belly, lower fins and deeply-forked tail are yellowish. The head is rounded, with a small mouth. A blue patch appears just above the eye. Pompano often leap out of the water and slap the tops of the waves. *Average size:* 1 to 3 pounds. *Maximum size:* 5 pounds. *Season:* Year-round; most common in cooler weather. *Where found:* Bay and inshore, the Gulf side of the causeway islands. *Bait:* Shrimp, sand fleas, small jigs. *Edibility:* Gourmet-quality, fine-textured meat.

Seatrout

Florida's most caught game fish, the **spotted seatrout** *(Cynoscion nebulosus)* will take any kind of lure, forgive a bad cast, show up year-round, and make for great eating. This fast, aggressive fish will hit your line decisively and fight all the way to the boat. The most brilliantly colored of all seatrout, the spotted seatrout has black dots peppering its upper sides, back, tail and second dorsal fin. The underside has a sky-blue tint. The mouth is soft and tears easily. There are usually two distinct canine teeth at the tip of the upper jaw. The sides are silvery, the back grayish to dark. Variations occur depending on the waters from which the fish is taken — darker from mud flats and back bays, lighter from

Cooked to order. The Lazy Flamingo restaurants will cook your catch (assuming you clean it first). Expect to pay $7-8 for 10 ounces.

Mullet leap from the water often, apparently to clean their gills or rid their bodies of parasites

Mullet roe is exported to Japan, where it sells for up to $100 a pound. Smoked mullet is popular with Florida natives.

Florida pompano brings the highest commercial price of any fish in Florida

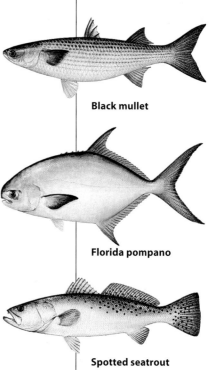

Black mullet

Florida pompano

Spotted seatrout

The Sanibel lighthouse fishing pier on San Carlos Bay. Below, busy lines bring big smiles.

inlets and channels. *Average size:* 2 to 3 pounds. *Maximum size:* 10 pounds. *Season:* Year-round, peak activity is spring and fall. *Where found:* Creeks, grass flats, the pier at the lighthouse, the bay and Gulf sides of the causeway islands, Wildlife Drive in the refuge, edges of oyster bars, other grassy bottom areas near sandy depressions. *Bait:* Jigs, live bait, lures. *Edibility:* Good to excellent, best during cooler months. Meat softens and deteriorates with long icing.

Sheepshead

A **sheepshead** *(Archosargus probatocephalus)* is easy to recognize by its chunky body outline and sharply contrasting light and dark vertical bars. In younger and more brightly colored fish, the stripes look like prison clothes. The mouth and incisor-like teeth — which can cut fishing line — resemble those of a sheep. Sheepshead are expert bait-stealers. With shrimp, they will hover near the bait, examine it, then quickly suck the meat out of the shell. You'll feel only a slight twitch. If you use a whole shrimp, the fish will take all but where the hook went into the meat. Use half a shrimp and it will take the meat from the open end of the shrimp. Use a piece without the shell, and you'll never even feel the bite.

Difficult to clean, sheepshead have a heavy set of ribs in their "shoulders" that make filleting tough. Some say the only

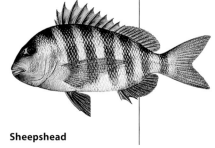

Sheepshead

way to clean a sheepshead is with a chain saw! *Average size:* 1 to 2 pounds. *Maximum size:* 11 pounds. *Season:* Winter. A few stay all year. *Where found:* Gulf, bay, the pier at the lighthouse, the bay beach and causeway islands. *Bait:* Shrimp, fiddler crabs, sand fleas. *Edibility:* Good baked or filleted, but skin first.

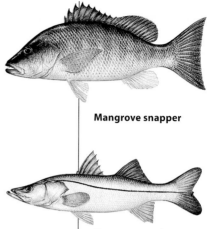

Mangrove snapper

Snapper

Reddish bands mark the sides of the **mangrove (gray) snapper** *(Lutjanus griseus).* The base color varies, depending on where it's caught (gray offshore, reddish in the bay). Dorsal and tail fins are dark; pectoral and anal fins are slightly pink. The eye is red. In young fish, a dark streak runs from the nose, across the eye, to the dorsal fin. Once landed, a snapper will actually snap at you, and latch onto a finger if given a chance. They also use an inhaling, snapping action when feeding. *Average size:* 1 to 2 pounds inshore, larger offshore. *Maximum size:* 10 pounds. *Season:* Year-round. *Where found:* Bay, Wildlife Drive. *Bait:* Lures, shrimp, minnows. *Edibility:* Great filleted.

Common snook

Size matters. Here are the maximum sizes (in pounds) of local fish:

- Black sea bass: 5
- Black mullet: 5
- Florida pompano: 5
- Gafftopsail catfish: 8
- Spotted seatrout: 10
- Mangrove snapper: 10
- Sheepshead: 11
- Red grouper: 20
- Spanish mackerel: 20
- Redfish: 30
- Kingfish: 40
- Snook: 40
- Black drum: 50
- Black grouper: 50
- Cobia: 70
- Tarpon: 250
- Goliath grouper: 700

Snook

Snook *(Centropomus undecimalis)* are the most pursued, but least caught, game fish in Florida waters. These tropical fish behave like a largemouth bass, but are much faster and more powerful. Some of the smartest fish around, snook are wary, shy, tricky and finicky. Experts in the art of escape, snook learn their environment quickly and will use oyster-covered mangrove roots, pilings, oyster bars, and even their own razor-sharp gill plates to cut your line and free themselves. Their abrasive mouths can quickly wear through the toughest of leader materials. They will leap in the air, and are masters at throwing the hook during the jump. Called "linesides" by some, "robalo" by others, the snook has an obvious black stripe on each side of its body, running from the top of the gill covers to the tail. The back is yellowish-brown, the sides and belly are yellowish with gold or brown-tipped fins. A long-bodied fish, the snook has a pointed snout and a protruding lower jaw. *Average size:* 8 to 12 pounds. *Maximum size:* 40 pounds. *Season:* All year. Summer is closed season for keeping snook (it's spawning season), but it's the best time for catch-and-release fishing. Snook become lethargic in cool temperatures. *Where found:* Pine Island

Bumpersticker on Captiva

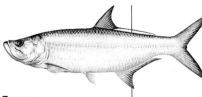

Tarpon

Unlike any other fish, tarpon can gulp air. They rise to the surface, roll, and breathe. On a still morning just at sunrise, you can hear them coming.

Tastes great. Eating your catch? Here's a guide to the best-tasting local fish:

■ **Gafftopsail catfish.** The meat is sweet.

■ **Cobia.** Meat from the smaller fish resembles grouper.

■ **Grouper.** Black and red grouper are excellent (goliath is protected). Red is the restaurant standard.

■ **King mackerel.** Try it filleted or cut as steaks.

■ **Mullet.** Great smoked.

■ **Pompano.** Top of the line. Gourmet quality.

■ **Redfish.** Blackened version made famous by Paul Prudhomme of New Orleans.

■ **Sea bass.** Good baked.

■ **Seatrout.** Best during cooler months.

■ **Sheepshead.** Good baked or filleted.

■ **Snapper.** Try it for breakfast.

■ **Snook.** Baked snook is excellent if skinned first.

Sound offers some of the best snook fishing to be had anywhere. Good spots include the pier at the lighthouse, the bay beach and causeway islands, Wildlife Drive and the mouth of the Caloosahatchee River and other pockets of deep water surrounded by shallow water, or bottlenecks between small mangrove islands where bait fish congregate. Many fishermen catch snook in the shade of mangrove trees. *Bait:* Shrimp, lures, cut bait, minnows. *Edibility:* Excellent up to 10 pounds. Larger fish tend to be coarse-textured, with less flavor. Snook should be filleted and skinned as the skin imparts a soapy flavor to the meat when cooked (snook used to be called "soapfish"). Properly prepared, fresh baked snook is delicious.

Tarpon

No other fish in Florida is as thrilling to catch as a **tarpon** (*Megalops atlanticus*). No fish jumps as high, is as hard to set a hook in or starts hearts pumping as fast. Once pricked with a hook, this great fighter immediately takes to the air, its huge gills extended and rattling, smashing off across the surface in leap after leap. This battle is easy to see, as tarpon often strike within 40 feet of your boat.

And you don't have to have a million-dollar boat or solid-gold reels to catch one. With light- to medium-weight tackle, a 14- to 25-foot boat and reasonable weather, tarpon can be caught by most anglers. Most use a flats boat, with a guide poling while the guide and the caster watch for fish. You can also catch one straight off the beach.

But you need to know what you're doing. Novice tarpon anglers can end up so frustrated they give up fishing.

A tarpon's hard, bony mouth makes it tough to set a hook. And when it senses danger, a tarpon can react so violently — a skilled, paced violence — keeping it on the line can seem impossible.

Tarpon migrate up to the area from the Keys in mid-April, hanging around the islands to feed in the shallows and flats until mid-summer. They swim past the beach, sometimes just five feet offshore, from the Sanibel lighthouse to Captiva Pass. Boca Grande Pass, just north of Captiva, is the acknowledged tarpon capital of the world. There's no other place on earth where tarpon gather in such large numbers to feed.

The tarpon is a heavy-bodied fish with a protruding, mean-looking lower jaw and huge mouth that points upward. A feather-like filament streams from the back of the dorsal fin. The scales are large, with the lower portion of each thickly coated in bright silver and the upper portion semitransparent. Tarpon around the islands are usually 3 to 7 feet long. They grow slowly: a fish over 100 pounds is at least 15 years old.

Average size: 50 to 200 pounds. *Maximum size:* 250 pounds. *Season:* Spring through summer, with peak activity in June and July. *Where found:* Bay, Gulf, the mouth of the Caloosahatchee River, Boca Grande Pass. *Bait:* Live or cut bait, chum, lures. *Edibility:* Not eaten in the U.S. Meat is dark, soft, lacking in flavor. Catch-and-release this game fish to provide thrills for others. Use a release gaff.

Fishing charters

There are dozens of fishing charters available on Sanibel and Captiva. Using a guide can be expensive, but convenient — the bait and equipment are provided, you don't need a license, and the guide will even clean your catch. And, of course, guides know where to fish, making it easy to have a great day. Half- and full-day charters are common. Some guides offer trips as short as two hours. If you have a good trip, your guide will expect to be tipped. Local guides include:

Back Country Fishing Charters *(433-1007)* with Capt. Paul Hobby, a lifetime Southwest Florida resident guide specializing in light tackle fishing for snook, redfish, tarpon and others in calm, shallow waters. Half-day and full-day trips are available.

Bait Box Guide Service *(1041 Periwinkle Way, Sanibel; 472-1618 or 472-3035)* offers over 100 years of local experience (Woodring family members have been guides here since the late 1800s), with charters for tarpon, snook, redfish, grouper, trout, snapper, cobia, shark and mackerel.

Bayside Charters *(454-4270)* with Capt. Kelly Kaminski provides 4-, 6- and 8-hour backwater fishing trips for trout, redfish, tarpon and snook.

Capt. Bob Sabatino *(at Jensen's Marina, Captiva; 472-1451 or 851-0330)* has been a fishing guide on the islands since 1960, and he's known for his experience, knowledge and dependability. Sabatino uses a 25-foot center console boat. He has been featured on "Fishing with Orlando Wilson."

Capt. Joe Burnsed's Charters *(at Castaways Marina, Sanibel; 472-8658)* provides back bay, deep-sea and fly-fishing. Half-day and full-day trips are available, as well as split and private trips.

Capt. Ozzie Lessinger *(at McCarthy's Marina, Captiva; 472-9074)* specializes in fly and light tackle fishing aboard a 17-foot tournament-rigged flats boat.

Tarpon can jump high enough to clear the top of a small boat and cover a distance of up to 20 feet. It will jump up to a dozen times during a catch.

Tarpon success is measured in the number of fish jumped, not caught. If you jump half a dozen tarpon in a morning, you've had a great day. A typical angler will be lucky to land one or two fish out of 10 hookups. An experienced fisherman, with good tackle, can reel in a 150-pounder in 25 to 30 minutes.

Tarpon tips

■ **Fish over a light bottom** to spot tarpon easily. Their dark bodies will look like silhouettes.

■ **During the day,** the best times can be during the first two hours of an incoming tide or the end of an outgoing tide

■ **You can catch tarpon at night, too,** especially in calm surface water with a light wind

Get a trophy of your tarpon catch by taking its picture at the side of the boat, then gently pulling a scale from the middle of the fish. The scale, which can be two inches across, can be dried and "mounted" on a plaque with the photo and details of the catch.

Island fishing secrets

■ **Keep your hands clean** when you handle bait. Fish can detect the odors of suntan lotion, cigarette smoke, gas and oil. If your hands get dirty, use lime juice or fish slime to eliminate these scents. Spitting on bait helps, too.

■ **Find the birds.** Wading birds strolling along the shore or standing by a pass indicate the presence of bait fish and, therefore, the fish that eat them, such as snook, redfish and trout. Offshore, noisy terns often fly above mackerel and other migratory fish. Pelicans dive above schools of shiners.

■ **Fish the waters that are slightly murky.** Crystal clear water may look nice in the brochures and TV ads, but it's not the best for fishing. Most fish need some sand and silt in the water to strike. Ideal visibility is 5 or 6 feet.

■ **Wear polarized sunglasses.** They reduce surface glare, helping you see into the water and avoid manatees and grass.

■ **Read your charts.** Much of Pine Island Sound is only a few feet deep, even at high tide. Respect your charts so you don't run aground.

■ **Walk it out.** If you run aground, turn off the engine, tilt it up, and pole or walk your way out to deep water.

Capt. Pat Lovetro *(at Sanibel Marina; 472-2723 or 826-3156)* has been fishing local waters for 20 years. Take out his 25-foot Parker for 4-hour, 6-hour or full-day trips to search for tarpon, shark, trout, snook and more.

Capt. Randy's Fishy Business Charters *(at 'Tween Waters Marina, Captiva; 472-2628)* features back bay or offshore fishing. Trips can also include shell collecting, dolphin watches, snorkeling, and breakfast or lunch trips on North Captiva, Cabbage Key, Useppa and Boca Grande. Child friendly.

Fintastic Charters *(850-0716)* and Capt. Randy Eastvold offers half-, three-quarters- and full-day trips into Pine Island Sound and the Gulf of Mexico for snook, redfish, tarpon, trout and shark. Shelling and sightseeing trips are also available.

Joyce Rehr's Fly Fishing & Light Tackle Guide Service *(1155 Buttonwood Ln., Sanibel; 472-3308)* will take you in the flats, back bay and offshore. Fly-fishing and light tackle. Novice to advanced anglers. Half-day, 3/4-day and full-day trips.

Makin' Waves Charter Services *(Captiva; 395-7696 or 560-6300)* and Capt. Tim Hickey provide either an 18-foot Hewes flats boat or a 26-foot Dusky — your choice. Half- or full-day trips for tarpon, snook, redfish, snapper and shark.

Santiva Saltwater Fishing Team *(at 'Tween Waters Marina, Captiva; 472-1779)* offers back bay fishing with native guides Capts. Jim and Jimmy Burnsed. Private 2-, 3- and 4-hour trips available. Children are welcome.

Sportfishing Charters *(Sanibel; 472-6089 or 770-7375)* features luxury fishing cruises for up to six aboard an air-conditioned 36-foot sportfishing yacht. Go deep-sea trolling or bottom fishing in up to 100-foot depths in the Gulf of Mexico for mahi-mahi, kingfish, wahoo, grouper, sailfish, snapper, barracuda, shark, tuna, red drum, tripletail and tarpon. Longer trips, overnight trips and diving trips are also available.

Tarpon Bay Recreation *(in the J.N. "Ding" Darling National Wildlife Refuge, 900 Tarpon Bay Rd., Sanibel; 472-8900)* offers guided canoe fishing tours. It also rents broad-beam and square-stern fishing canoes, with electric trolling motors. Bait and tackle are available in the rental store.

Ultimate Charters *(542-9315)* with Capt. Gary Clark provides half-day or full-day trips for snook, tarpon, cobia, trout and others. Gary has two new tournament-equipped 20-foot back-country flats boats and state-of-the-art light tackle. Fly tackle is also available, or you can bring your own.

Catch and Release

Today most fishermen in Florida don't keep their catch. They release many, if not most, of their fish back in the water so they can continue to reproduce. Regulated species, such as redfish, should be released if they are outside their slot limit, either too small or too large.

Some anglers release everything they catch. But it helps to know how to do it.

Taking an exhausted fish out of the water is like placing a plastic bag over the head of a marathon runner. It needs oxygen! Fish that are in good shape should be released immediately by removing the hook, or cutting the leader as close to the hook as possible.

Large gamefish such as billfish, tuna, shark and tarpon (above) should be brought alongside the boat as quickly as possible. Don't boat these big guys — they're dangerous to both themselves and you. Many times large gamefish evert their stomachs when hooked. Don't attempt to replace it; the fish will swallow it after release.

Here are a few other catch-and-release tips:

1. Don't wear it out. Playing the fish to exhaustion depletes its energy reserves, which puts it at risk of death by metabolic imbalances and makes it easy prey.

2. Use barbless hooks. Using circle hooks with natural baits reduces gut hooking and increases your hookup ratio. Don't use hooks coated with cadmium, which is toxic to fish.

3. Use artificial lures when possible. Fishing with artificial bait, with single hooks, decreases the likelihood of gut-hooking.

4. Avoid gut hooking. Set the hook immediately on the strike, before the fish swallows the bait. If you do gut-hook a fish, cut the line and leave the hook in place. Don't lift gut-hooked fish by the leader; this increases tissue damage.

5. Wet your hands. If you can't leave the fish in the water during a release, gently cradle it under the rib cage using wet hands. Nets, dry hands, towels and gloves remove too much of the fish's protective slime and put it at risk of infection.

6. Handle the fish as little as possible. Never lift a fish by the gills or eyes if it is to be released. A fish can be calmed during release-handling by turning it on its back or by covering its eyes with a wet towel. Use needle-nose pliers or a dehooker to remove the hook. Keep these tools in a convenient place so you can release fish quickly.

7. Revive the fish. Hold the fish head-first into the current until it swims away. Don't throw the fish into the water, or drop it sideways. If the fish is exhausted, revive it by making sure the head is totally submerged and towing it gently forward. When releasing a fish from a bridge, pier or boat, gently drop it into the water head-first to reduce the impact and force water through its gills.

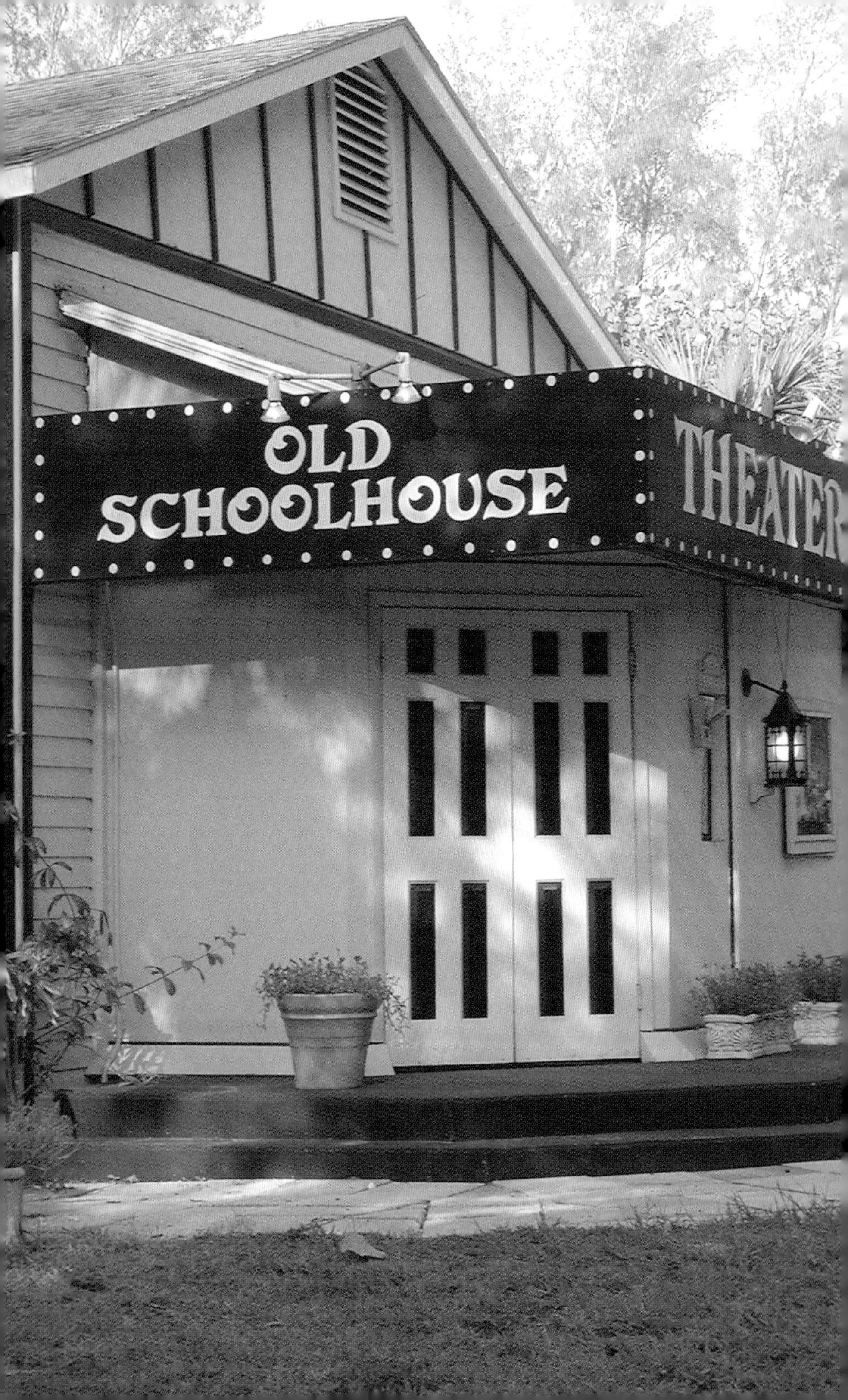
OLD SCHOOLHOUSE
THEATER

Diversions

*W*hat do you do when you've had your fill of beaches and birds, fauna and fish? First, consider the Bailey-Matthews Shell Museum (page 118) and the Sanibel Historical Village and Museum (page 65). The islands offer a variety of other activities, too, including live theater, golf and a first-class library. More unique diversions are just a short drive away on the Florida mainland.

Live theater

The casual and friendly **Old Schoolhouse Theater** *(1905 Periwinkle Way, Sanibel; 472-6862)* is one of the island's best bets for a fun evening out. The building itself is Sanibel's old one-room schoolhouse dating to 1896. Broadway veteran Ruth Hunter transformed it into a tiny nonprofit theater in 1964. For the past 10 years it has been home to artist-in-residence J.T. Smith and his high-energy, original musicals. Shows get a little randy at times, but are generally appropriate for children. The Christmas "Ho Ho Ho" show is an island tradition.

The "L"-shaped seating area holds only 92 people; the farthest seat from the stage is only seven rows back. The theater recently installed new padded seats. Volunteers offer beer, soft drinks and snacks at intermission on a pay-what-you-feel basis. Many actors who perform here go on to Broadway and network television roles. Showtime is 8 p.m. Monday through Saturday. Occasional matinees are Wednesdays at 2 p.m. and Saturdays at 4 p.m. Tickets are $25.

Formerly the Pirate Playhouse, the **J. Howard Wood Theatre** *(2200 Periwinkle Way, Sanibel; 472-0006)* was recently renamed for the late publisher of the Chicago Tribune, a Sanibel winter resident and island benefactor who made a large donation to the nonprofit group. It hosts comedies, dramas, classics and Shakespearean plays. Producing director Robert Schelhammer took over here during a controversial change of staff two years ago. Performances are at 8 p.m. Monday through Saturday during the season, with a matinee Wednesday at 2 p.m. The first Friday of a run is opening night, with a reception at 7 p.m. The second Tuesday has post-show discussions with the actors and director. Special programs include the family-oriented Pirate Players Children's Theatre, often with island children in the cast (tickets are $5, call for times) and the Play Reading Series and Playwright Festival, where attendees read a play together, then discuss it with the director ($10, Sundays at 7 p.m.) Regular performances are $25 for adults and $15 for students, plus a $2 handling charge. Children 4 and under are free. Discounts are available for groups of 20 or more. Tickets have to be purchased 48 hours before show time. The 180-seat theater sits in an impressive building designed as a theater from the ground up. Free backstage tours are available.

Facing page: Sanibel's Old Schoolhouse Theater building was a school from 1896 to 1963

Sanibel's Wood Theatre is the island's "serious" playhouse. It hosts comedies, dramas, classics and Shakespearean plays.

Beachview Golf Course is the easier of the two Sanibel public courses

Golf and tennis

Pay attention when you play golf on the islands. You never know when you'll walk up on a snake, alligator or bobcat. "My buddy and I were out at the Dunes one day, riding along in the cart, talking away, not paying much attention," says islander Brad Sitton. "Next thing I know, I look up and there's a 10-foot gator staring me in the face. Nearly ran right over him." Golfers of every level can have a good time here. Fairways and greens are Bermuda grass. Most have been built with the leisurely duffer in mind.

The par-70, 5,600-yard course at the semi-private **Dunes Golf & Tennis Club** *(949 Sandcastle Rd., Sanibel; 472-2535)* has water on all 18 holes. Most greens have forced carries over water; one has an island fairway. You'll need to hit your mid-to-low irons consistently well. Touring pro Mark McCumber redesigned the course in 1995. The new layout has an emphasis on chipping. The Dunes also has a water range, the only one on the islands. The clubhouse includes a dining room, grill and bar. Club rental, repair, fitting and PGA instruction are available. Pro shop hours are 7 a.m. to 5:30 p.m. daily. The pro is Kevin McCune. Make reservations at least four days in advance (six days for South Seas Resorts guests). For tennis, the Dunes has seven soft clay Har-Tru courts. There's a summer tennis camp for kids. The tennis pro is Terry See.

The semiprivate course at the **Beachview Golf & Tennis Club** *(1100 Parview Dr., Sanibel; 472-2626)* has numerous waterways. Older players and duffers who play once or twice a year will feel at home on this 6,320-yard, par-71 course. The fairways are open and the greens are generous. Coconut and cabbage palms, as well as macadamia, grapefruit and lime trees, are common throughout the course. The clubhouse includes a snack bar, restaurant and lounge. Club rental, repair and fitting are available. Pro shop hours are 7 a.m. to 5:30 p.m. daily. Pro Robby Wilson has been here since 1974. Call at least two days in advance for reservations, earlier during season. Club members have priority. The Beachview tennis center features five HydroGrid clay courts. They're watered from underneath, so they're rarely closed and stay in good shape. The surrounding facility has an elevated deck with 270 degrees of courtside viewing. Daily packages are available. The tennis pro shop is open 8 a.m. to 5:30 p.m. daily. Private lessons are available from pro Russ Crutchfield.

The **South Seas Resort** on Captiva has a 9-hole course for its guests and members, with fairways along the beach. Sanibel is also home to the ultra-exclusive **Sanctuary Golf Club,** with a beautiful course that winds through palm hammocks and along the bay. No one ever, ever, gets on the links here except members and their guests.

USPGA course ratings

Dunes

Course/slope		
Back	68.4	124
Middle	67.0	121
Ladies	64.9	113

Beachview

Course/slope		
Blue	70.8	127
White	67.8	118
Gold	63.5	105
Red	67.6	114

Children have free fun at the Sanibel Recreational Complex

Fitness centers

With the Sanibel School located right in the middle, most visitors (and many residents) assume the **Sanibel Recreational Complex** *(3840 Sanibel-Captiva Rd., Sanibel; 472-0345)* is only for local kids. But everyone is welcome here. In fact, nonresidents make up 30 percent of the crowd. The free facility is run by the city of Sanibel, and features a large pool with swimming lanes; an indoor gym with basketball courts and a weight room; and baseball, tennis and volleyball courts. (The outdoor courts are on the right side of the Sanibel School. There are no reservations; you just show up and play.) A large playground behind the gym has swing sets, picnic tables and basketball hoops. A huge wooden play area is a maze of nooks and crannies. A new smaller play area for younger kids was built in 2000, between the ball fields.

Attend an NPR concert. The Sanibel Music Festival (336-7999) offers a series of performances each spring at the Sanibel Congregational Church, some broadcast on National Public Radio. A true delight, the festival includes piano and chamber music, two operas and a children's concert. Artists have included the Julliard String Quartet. Ticket prices are low for shows of this quality: $22 to $30.

See a movie. Probably the smallest movie theater you'll ever be in, the two-screen Island Cinema (next to Bailey's General Store, Sanibel; 472-1701) is newly renovated. It shows the latest releases.

Hear songs of Sanibel and Captiva from musician Danny Morgan. He performs on tour boats and at special island events throughout the year.

The exercise classes at the newly expanded **Sanibel Fitness Center** *(975 Rabbit Rd., Sanibel; 395-BODY)* include aerobics, body-shaping and indoor cycling. Other classes feature yoga, karate and weight-loss, and exercise and dance for kids. Equipment includes treadmills, Stairmasters, bikes, free weights and hi-tech Cybex machines. Certified personal trainers are available for cardiac rehabilitation and other special needs. Nonmembers are welcome for a fee of $10 a day or $40 for six days in a month. Monthly fees start at $50. The center is open Monday through Saturday year-round, and on Sundays during the season.

The historic **Sanibel Community House** *(2173 Periwinkle Way, Sanibel; 472-2155)* is the setting for a series of fitness programs sponsored by the Sanibel Community Association. Typical classes include Kripalu Hatha Yoga, Pilates mat classes, no-impact workouts and Fitness with Carla, a cardiovascular workout balanced with stretching and muscle toning. (We can vouch for Carla Ferrel's effectiveness. She taught our daughter to swim.) Visitors are welcome. The guest fee is $7 per class.

Administered by the city of Sanibel, the **Island Seniors Program** at the Sanibel Civic Center *(2401 Library Way, Sanibel; 472-5743)* is designed for people over 50. Activities include low-impact aerobics, line dancing, bowling, hiking and swimming. Carla is one of the instructors here, too. Lifeguard Coby Amadio conducts water aerobics at the Sanibel Recreation Center. You have to be a member to participate here, but the yearly fee is only $15.

BIG Arts

The **Barrier Island Group for the Arts** *(between Sanibel City Hall and the Sanibel Historical Village, 900 Dunlop Rd., Sanibel; 395-0900)* has a performance hall, two art galleries and five classrooms. About two dozen jazz, classical and pop concerts are held each year in the 414-seat Schein Performance Hall, with artists from throughout the U.S. and Europe. The Phillips Gallery has exhibits and juried shows which change monthly, with all art for sale (hours are generally 1 to 4 p.m. weekdays). Over 160 workshops teach painting, pottery, weaving, dance, photography and other skills. Most are available on a persession basis, making it easy for visitors to attend. Discussion groups meet to talk about books, current events, music and art. The BIG Arts Film Society brings residents and visitors together to view and discuss selected movies, with a focus on independent and foreign films that have not been widely seen in commercial theaters.

Other activities include lectures, language workshops,

civic events, a Thanksgiving-weekend juried Fine Arts and Contemporary Craft Fair, fitness classes including Pilates body conditioning, yoga and dance, and programs for children. A hub of island cultural activity, BIG Arts was established in 1979. It's closed on holidays.

Libraries

Open to residents and visitors, the **Sanibel Public Library** *(770 Dunlop Rd., Sanibel; 472-2483)* has over 50,000 books. They include multiple copies of current best-sellers, a local-history collection and numerous videos, CDs and books on tape, as well as a great children's section. The impressive 19,000-square-foot building has many clean, comfortable spots to read and relax. The lobby has collections of local shells and fossils, including a tooth of a prehistoric North American camel. The library offers free high-speed Internet access, available to everyone (even those without a library card). Programs include "Stories for Grown-ups" the first Friday of each month, book discussion groups, lectures and book signings. Children's activities include "Books for Babies," preschool story hours, "Young Puppeteers" and holiday events. Nonresidents can get a library card for $10, good for one year. The library is closed on Sundays, but open late (til 8 p.m.) on Mondays and Thursdays.

The tiny **Captiva Memorial Library** *(11560 Chapin Lane, Captiva; 472-2133)* has books of local interest, shelling and nature guides, magazines, videos and books on tape, as well as Internet access and a word processor. Hours are Tuesday, Thursday, Friday and Saturday 9 a.m. to 5 p.m.; Wednesdays noon to 8 p.m. Visitor cards are $5.

Day trips

The **Edison-Ford Winter Estates** preserve 1920s-era seasonal homes of Thomas Edison and next-door neighbor Henry Ford *(2350 McGregor Blvd., Fort Myers; 334-3614, 888-377-9475)*, nestled on the banks of the Caloosahatchee River. Daily 90-minute tours combine both estates. The main tour includes Edison's gardens, the world's largest collection of his inventions and memorabilia, a stop

The Sanibel Public Library began in a closet-sized room in the Sanibel Community House in the 1960s. Volunteers collected books left in hotel rooms, and donated others from their homes. Today's building (top) was built with private donations. Unexpected touches include a reading porch and murals in the children's library that depict island-style Mother Goose rhymes. Above, the sea cow (manatee) jumps over the moon.

Edison home

Sun Harvest Citrus

Below and facing page:
Have a beer and watch
the Red Sox in Fort Myers

at his winter laboratory, and a few of Ford's early automobiles. If you've got the time, check out the longer historical or botanical tour, or the super trip that includes a ride through the Fort Myers historic district and a boat ride on the Caloosahatchee. Reservations are required for the longer tours. Open daily except Thanksgiving and Christmas.

Get an up-close look at Florida's citrus industry at **Sun Harvest Citrus.** It offers free tours of its 24,000-square-foot packing-house *(14810 Metro Pkwy., Fort Myers; 768-2686 or 800-743-1480).* A cafe offers freshly-prepared sandwiches and orange and lime ice cream. The adjacent market has gourmet foods and free juice samples. Groups of more than 10 require reservations, otherwise you can just drop in.

The **Koreshan State Historic Site** *(U.S. 41 at Corkscrew Rd., Estero; 992-0311)* recognizes a defunct religious community from the late 1800s that believed the earth is a hollow sphere and that the planets, moon and sun exist at the core. Envisioned for 10 million residents, the group's planned Utopian city — preserved here — never got above 250. You can tour the grounds, or rent a canoe.

The National Audubon Society manages the 11,000-acre **Corkscrew Swamp Sanctuary** *(375 Sanctuary Rd., Naples; 348-9151).* The preserve features the country's largest stand of virgin bald cypress trees. Take the boardwalk tour of pinelands, wet prairies, hammocks and cypress ponds.

Spring training has been a tradition in Fort Myers since the 1920s. The home teams are the Boston Red Sox and Minnesota Twins, who each play about a half-dozen games here each March. The Red Sox hold their spring training games at the 6,900-seat City of Palms Park *(downtown at 2201 Edison Ave., Ft. Myers; 334-4700).* The Twins play at the 7,500-seat Lee County Sports Complex *(near the airport at 14100 Ben C. Pratt/Six Mile Cypress Pkwy., Ft. Myers; 338-9467).* The small stadiums give you a unique, up-close-and-personal experience. It's a great way to spend an afternoon, especially when major stars are playing. Tickets average $10.

Starry Starry Nights
Island stargazing

In a typical U.S. suburb you can see about 200 stars in the night sky. But on Sanibel and Captiva you can see 3,000. The skies are ideal for stargazing here; clear, smog-free, and pitch black. Much of the land, including all the beaches, is restricted from development, so there are no lights. Even commercial areas stay dark. Street lights have been banned, as have those mercury-vapor and high-pressure sodium lights so common elsewhere.

And the night air is usually calm, giving you a clear view of indistinct objects.

You can see plenty even without a telescope; just pull out your binoculars and look up. You'll have a terrific view of the moon's surface, the phases of Venus, the four large moons of Jupiter, even the polar caps of Mars.

Pick a clear night and head outside to the beach or other dark spot. Wait 10 minutes for your eyes to adapt, then start looking at the sky. To see a faint object more clearly look slightly to the side of it, letting its light fall on the more sensitive, outer part of your eye. Try to avoid pitch-black skies (the profusion of stars makes identification more difficult) and nights with a full moon (too few stars will be visible, and the direct sunlight minimizes shadows on the moon).

Many features on the **moon's** surface are easy to see on Sanibel and Captiva. With a good pair of binoculars you can spot dozens of craters, mountain peaks and even subtle ripples on the plains. The moon is highest in the sky here in December, when it passes directly overhead. In the summer it rises only about 40 degrees.

The **Milky Way galaxy** appears as a misty band arcing over Sanibel and Captiva during the summer. In winter it's dimmer and less structured. Our neighborhood in the universe, the Milky Way has several hundred billion stars, though you can see only 5 percent of them with the naked eye. Our sun is about two-thirds of the way out from the center.

Once you've spotted one of the **planets** it's easy to see the rest: they're always in a straight line! Called the ecliptic, this line includes the sun, moon and all the eight other planets. Mercury, Venus, Mars, Jupiter and Saturn are visible from the islands with the naked eye.

Mercury is rarely seen; its tight orbit keeps it close to the sun. It hugs the morning or evening horizon. The darkest planet, it reflects only 6 percent of the sunlight it receives.

Venus, the brightest planet, alternates between being our "morning star" and "evening star," depending on its orbit. It's visible either for four hours after sunset or for four hours before sunrise. Look for Venus near the horizon in the twilight.

Called the red planet because sunlight reflects off its reddish deserts, Mars often looks like a pale red dot. It changes brightness more than any other planet, as its distance to Earth varies greatly as it circles the sun. With a small telescope you can see the planet's polar caps. With a large telescope you can see its two moons.

Jupiter glows a steady, creamy white. The second brightest planet, its four largest moons can often be seen on the islands with binoculars. Also look for the shadow cast by a Jovian moon, looking like a bullet hole in Jupiter's surface. You may be able to see the Great Red Spot, a cloud system three times wider than the Earth.

You can see the rings of Saturn with even the smallest telescope, sometimes even a pair of binoculars. Also look for the shadows the rings cast on the planet's surface. Saturn itself is a pale yellow orb.

You can see Uranus (pronounced "YER-an-us," not "your-AY-nus") with binoculars, appearing as a small blue-green disk, as well as Neptune, though it's tough to identify. Pluto is impossible to see without at least a 6-inch telescope. (Junior-high trivia: The blue-green color of Uranus is from a covering of methane gas.)

How do you tell a planet from a star? Planets don't twinkle like stars do. Twinkling occurs due to atmospheric turbulence. Planets are close enough to Earth to look like tiny disks instead of pinpoints of wavering light.

All major **constellations,** including the Southern Cross, are relatively easy to see in the dark island skies. The seven stars of the Big Dipper look like a ladle. But they're actually part of a larger constellation, Ursa Major, the large bear in mythology that guards the polar regions. The Big Dipper is always above the horizon, never rising or setting. Its stars include "The Pointers" — the two stars on the far side of the bowl from the handle.

Orion ("oh-RYE-un") is the second most prominent constellation. It represents the great hunter who bragged that he could kill any creature on earth, but met his match when a scorpion stung him on his heel and killed him. The constellation has a glittering three-star belt framed by four prominent stars, set in an hourglass, marking the hunter's shoulders and legs.

Sanibel and Captiva are the northernmost areas to see the Southern Cross, the constellation Crux. In the spring look for it due south, just above the horizon. The best month to see it is April. The Southern Cross actually looks more like the "Southern Kite," as there is no bright central star. The smallest constellation, its vertical bar points to true south.

Each **star** is a sun, just like our own. But the stars vary greatly in size and luminosity. And, from our vantage point, everything depends upon the star's distance from Earth: The closer stars look brighter, the most distant stars are invisible. Marking true north, Polaris ("poh-LAIR-iss") is the Night Star, one of the brightest stars in the sky. Find it by looking at the Big Dipper's "Pointers," and extending an imaginary line upward. Sirius ("SEAR-ee-us," which means "scorching" in Greek) is the brightest star in sky, even brighter than most planets. Only Jupiter, Venus and Mars are brighter. Locate Sirius by first finding Orion's belt, then following the imaginary line of the belt to the left. When the air is turbulent, Sirius appears to twinkle violently, changing colors from white to blue to yellow, glittering like a diamond. When the air is still, the star has a bluish-white tinge.

Each night you can see many **satellites** in the island sky, including the International Space Station, the Hubble Space Telescope, the Tropical Rainfall Monitoring Mission and many spent Cosmos, Milstar, Delta, Progress and Meteor rockets. These satellites take just two or three minutes to cross the sky. The best time to look is during the first hour of darkness in the spring and summer, or the last hour before sunrise, as most satellites are only high enough to catch and reflect sunlight during these times. A careful observer on the islands should see at least 10 satellites in the first hour after nightfall.

How do you tell a satellite from an airplane? Satellites fade quickly from view in a clear sky; a high-flying plane doesn't. Also, satellites appear white, like stars without the twinkles. Most airplanes have flashing lights. For up-to-date, local information on satellite viewing times visit www.heavens-above.com. (Don't have your PC with you? Surf online for free at the Sanibel Public Library.)

You can even watch the **Space Shuttle** take off from the islands. The best viewing spot is the Sanibel fishing pier or the Sanibel Causeway. Look to the northeast to see the smoke trail arc up left to right across the sky.

Sanibel was the first Florida community to adopt a Dark Skies ordinance, forcing all lights on the island to point down. The only exceptions are Christmas lights and the lighthouse beacon. The law is being phased in over 15 years to give businesses time to retrofit commercial signs.

Island Living

By Mike Neal

*L*iving on Sanibel is living in paradise. You go to the beach whenever you want, and wear T-shirts and shorts every day. You have key lime pie for dessert; coconut milk when you're thirsty. You have a pet dolphin, just like Flipper. And you live in a thatched hut, just like Gilligan and the Skipper.

At least that's what we thought, living in Atlanta watching perhaps a little too much Nick at Nite. Then we moved here, on the 4th of July, 1996. Driving the U-Haul all night from Georgia, we pulled into the driveway of our new home at dawn. We had done it — moved to an island! We sat in our vacant dining room eating convenience-store doughnuts and orange juice, and watched the sun come up.

The next morning I took our daughter to the Sanibel pool. I pulled onto San-Cap and accelerated to my customary 65 mph.

"Wonder who they're after," I thought to myself, as I saw the red lights in my rear-view mirror. After being politely lectured on the island's 35 mph speed limits ("You're kidding!" I said, which didn't help), I drove on to the pool. Micaela and I hopped in the water, and *a woman I did not know* came up and started a conversation. "I'm Lisa Williams," the stranger said. Beat. Oh, my turn. "Mike Neal, and this is Micaela." It was the first time I had introduced myself to a stranger, *just right out in public,* in at least a decade. Later I drove up to the Hess station to buy gas, and — *voilá* — "Hi Mike" — there's *the same woman again,* pumping gas next to me! Now what do I say? What was her name? How does this work? *Why is she following me?* Then I drove over to the post office and… "Hi Mike." There she was again!

Never, ever had I lived anywhere where people actually drive 35 mph, or where you see people you know over and over again, the *same* people in the *same* day. "This is bizarre," I told Julie when I got back home. "I think we've moved to Mayberry."

And that's just what Sanibel is: a small town on a slow pace, where everyone knows everyone else. "I have never been to Bailey's without seeing someone I know," says 12-year resident Melissa Congress, who helps run her family's jewelry store. "It's not possible to be a loner," adds Charlotte Harlow, an island homemaker who helps raise money for the Sanibel School. Community spirit runs high. "When I had a baby people made me food, people I didn't know that well," Melissa says.

Islanders feel safe. Our kids roam freely on the bike paths; some ride their bikes to school. Many of us don't lock our cars, or our homes. And we're certainly not afraid of the dark — we take long walks at night on the pitch-black beaches.

It's also a subtropical, island life. Living here is like being on a vacation that never ends

---

Facing page: A Sanibel teenager tosses flags to the crowd at the annual 4th of July parade

All smiles: Sanibel's Rudy and Sandy Zahorchak

The drinking water on Sanibel and Captiva is almost too good — literally. Created through a reverse-osmosis system (the same method as most bottled water), island water is almost completely deionized. The Island Water Association actually adds some hardness into it to protect residents' pipes and hot water heaters. The water comes from an underground river 750 feet below the surface. The islands use 3.5 million gallons of water a day; 25 percent goes for pools and landscaping.

Facing page: A hobbyist makes shell caps at the Sanibel Shell Fair

— you never go back to the real world. Yes, you do go to the beach whenever you want, and your wardrobe is almost exclusively T-shirts and shorts. "People wear their weekend clothes every day, shoes are optional," Charlotte says. "My husband's goal is to wear shorts 365 days a year. So far he's made it to 363." Melissa agrees. "You only wear socks when you go off the island," she adds. "And my husband hasn't bought a suit since we got married."

You forget about shock jocks, talk radio and pop culture. Many residents — even, now, Julie and I — watch little TV. Instead you volunteer — at the refuge, SCCF, CROW, the school, the preschool. In your garage your cars fight for space with bicycles, kayaks, Waverunners and windsurfers. In your yard you grow limes, mangos and, yes, coconuts. By the way, coconut milk tastes *awful*.

And you stay happy. Surrounded by excited, appreciative tourists, you're always reminded how special it is here, and how fortunate you are to call these islands home.

Who lives here

Few islanders are native Floridians. For the most part we're simply Northern tourists, generally over 30, who fell in love with the place and became determined to move here. About 3,000 full-time residents live on the islands, 91 percent on Sanibel. And indeed, most of us moved here not for the climate, but for the community — often because it's such a great place to raise kids.

Sick of living in New Jersey, Rudy and Sandy

Thoughts While Waiting To Make a Left Turn onto Periwinkle Way at Height of Season

By Joseph Pacheco

Creating the paradise is easiest:
You pick a place everyone overlooked,
That goes against the current fashion,
Has some very special things: beaches
Filled with shells, a great big sanctuary
For animals, birds, trees and man
To be as close as they can ever be
And a panoramic bridge
    to get you there.

Then you pick out your spot
    and build on it
For far less the cost you hope
It will someday be worth,
And you revel in the restrictions and the
    limitations
You would never have put up with
    elsewhere:
Government telling people what to do
To keep other people from pouring in
And ruining your Eden.

Maintaining the paradise is harder:
The place is discovered, it's hot,
    everyone wants it,
The price of land and houses,
    everything, rises
Like the Australian Pines
    you want to topple;
Builders burst out of their woodwork

Building like beavers before a flood,
Demolishing and replacing
    the bungalows
Of those who loved paradise first —
The battle to contain them rages
And you choose the side of
    necessary ordinance
And the drawbridge dragon guarding
    the moat of your island.

Travel magazines put you
    on their covers
And lists of the 50 best places
    to live and to go,
Everyone's renting and buying
    and suddenly
There are never vacancies and
    the whole world,
It seems, wants to taste and drink
The milk of your paradise.

Leaving the paradise is hardest —
The anaconda of pick-up trucks,
    campers and convertibles
Slowly choking Periwinkle Way
Suddenly relaxes,
A tiny space appears
    between two SUVs
And like a wild animal set free,
You make your left turn.

Zahorchak were looking for a place to raise their new-born son, Rudy Jr., in 1995. At first they refused to consider Florida. "We thought it was too hot and full of old people," says Rudy, a pilot for Northwest Airlines. But then they came here for vacation, and discovered Sanibel's intimate culture. "I grew up in a small town in Wisconsin," Sandy says. "I knew everyone in my school, and if I went to the grocery store I'd always see someone I knew. It's like that on Sanibel. I like that little Rudy gets to grow up in a small town."

New York physician Steven Wener and his wife, Susan, had been vacationing on the islands for years when their son Matthew was born in 1993. "Long Island was no place to raise a child," Susan says. "It was so stressful, with a 90-minute commute to town." Determined to relocate to Sanibel, Steven gave up his private practice for a job in a Fort Myers veterans hospital — taking a 30 percent pay cut. "It was absolutely worth it," he says.

With no major job market nearby, many of us work out of our homes, connected to New York, Chicago or Miami by e-mail, fax and FedEx. Some run the islands' restaurants, shops and service companies. Others work as waitresses, teachers, lifeguards or police officers. Only a handful commute to the mainland.

Politically we're a strange mix. Conservative on many issues (most of us voted for Bush), islanders are also the strongest environmental voters in the state. And we diligently protect our paradise from overdevelopment, even raising our own taxes when necessary. In November, 2001, Sanibel residents voted to increase their property taxes to allow the city to buy up nine prime commercial acres along Periwinkle Way. "The alternative would have been a large shopping center," says Dick Walsh, Sanibel's vice mayor.

Mirroring the rest of Southwest Florida, Sanibel has many retired residents. Former Brooklyn Dodgers and Cincinnati Reds pitcher Jim Sprankle has reinvented himself as an avid saltwater fisherman and a nationally-famous waterfowl wood carver (some of his carvings hang in the visitors center of the J.N. "Ding" Darling National Wildlife Refuge). He's also the president of the Ding Darling Wildlife Society. Jim, his wife and young son moved to Sanibel in 1995, after visiting the island for years. "I just couldn't help picturing living here," he says.

Fifteen thousand seasonal residents dwarf the full-time population from October through March. Malcolm and Susan Harpham make the trek each fall from Connecticut. They volunteer at the refuge and the Chamber of Commerce and belong to the local fishing club, Power Squadron and Audubon Society. "We love it here," says

Housing for all.
Skyrocketing real-estate prices are great if you plan to sell (or if you're a real estate agent). But what if you're retired, on a fixed income? That was the problem on Sanibel in 1979. Many long-time, older residents were being forced to leave simply because they couldn't pay the property taxes. Sanibel had always been a community for all, not just the wealthy, and residents wanted to keep it that way. In response, island officials created Sanibel's Below Market Rate Housing program (BMRH). A collection of 50 city-owned small homes and apartments rent for as little as $445 a month; tenants include more than 50 children. Unlike most public housing, these units are located throughout the community and are indistinguishable from the "regular" homes next door. Prospective residents have to already live or work on the island and meet strict income limits: from $27,450 for an individual up to $47,450 for a family of five.

Same-day delivery.
Because the islands have few streets, a local letter mailed early in the morning at the Sanibel post office will be delivered that same day.

Susan, a former Harlem schoolteacher. "The people are so optimistic, so stimulating and interesting." They bought their Sanibel condo in 1998, on their first visit here. "We fell in love immediately — the beaches, the golfing, the biking, the refuge," says Malcolm, a retired steel-company executive. "We signed a contract before we flew back."

Education

The Sanibel School, a public K-8, stresses environmental education. Shell Studies are a part of the curriculum — kids study mollusks throughout the year then run the live-shell exhibit at the annual Sanibel Shell Fair. Some also work as docents at the adjacent J.N. "Ding" Darling National Wildlife Refuge. Many classes are held outside, often in the 26-acre Pick Preserve across the street.

Surrounded by the Sanibel Recreational Complex, the school also essentially has its own swimming pool, fitness center, soccer fields and tennis courts. The city of Sanibel sponsors after-school sports programs and a cheerleader squad; island parents run their own baseball and soccer leagues. Boy Scouts, Girl Scouts and church groups also thrive.

Many of the youngest islanders attend the remarkable Children's Education Center, a parent-owned preschool. About 60 kids paint, cook, play and sing with wide-eyed enthusiasm on the five-acre campus, which islanders have filled with everything from a state-of-the-art computer lab to a pedal-car race track. When we recently stopped by the kids were singing to each other — in English, Spanish, French, Japanese and *sign language.*

Government

The island of Sanibel is also the city of Sanibel. The city council often leads the country on environmental issues. Sanibel has the strictest tree-protection laws in Florida (all native trees are protected), has banned drive-through restaurants, and recently passed a Dark Skies ordinance that fights light pollution. "On Sanibel environmental issues are addressed first, before people issues," says Natural Resources Director Dr. Rob Loflin. "It's a model for how a community can be run," adds current Mayor Nola Theiss.

Captiva is an unincorporated area of Lee County. Its residents voice their concerns through the Captiva Civic Association.

Facing page: Jocelyn Harder, 7, bats for the Sani-Belles softball team

With an average height above sea level of only 4 feet, minor flooding is common on the islands, but hurricanes are the real threat. Hurricane Georges was headed directly for the islands in 1998; forecasters predicted storm surges up to 12 feet. But the storm changed course at the last second, sparing islanders from all but minor damage.

Sanibel is the only U.S. city that contains a national wildlife refuge within its boundaries

Sanibel police, such as officer Terri Cummins, use Service Oriented Policing, and answer every call

Sanibel Brownie Troop 641. Left to right: Bottom row: Micaela Neal, Nicole Horton, Lexie Dekker, Samantha Davies, Kari Fowler, Marin Williams, Nicole Ogden, Samantha Blaze. Middle Row: Shion Kremer, Caitlin Radigan, Elizabeth Sitton, Terry Earle, Jocelyn Harder, Julia Leal, Anna Aulino, Kayla Weber, Jessica Deforest. Back row: Troop leaders Lauren Davies, Karen Aulino, Lisa Williams.

The Passionate Retiree to His Beneficiary

By Joseph Pacheco

Retire with me to Sanibel
And we will spend our gold years well,
Preserving wetlands, beach and fields
With IRAs and pension yields;

Where sanctuaries free wild life
From human and survival strife,
Where CROW's not quite
　　the bird you think,
And 'gator aid is not a drink.

'Neath stars made bright
　　by true dark skies
Toward which no high-rise dare to rise,
We'll stand and guard a turtle nest
While you're decked out in Chico's best.

Perchance we'll catch a hatchling run
Or watch a mountain of orange sun
Like jelly on horizon spread
Drop swiftly into twilight's bed.

Midst morning walks upon the beach,
With myriad shells within our reach,
And skin-so-soft as insect spray
To keep no-see-um gnats at bay,

We'll stop and stoop and sift and crouch,
Put periwinkles in a pouch,
Collect sand dollars, cones and pens
To send up north to freezing friends.

Perhaps we'll find junonias rare
And learn which resort placed them there.
A necklace made of every shell —
Just live with me on Sanibel.

I'll clear all permits and build for you
A mega-house just made for two,
With full garage and space for three:
Our golf cart, Bimmer and SUV.

A forty footer to cruise the seas,
Catch snook and threaten manatees,
Play tag with dolphins as they leap,
With galley and two berths to sleep,

With latest radar and controls —
I'll join the power squad patrols,
Erect a dock to awe inspire
And cage our pool with finest wire.

When snowbirds crowd us in the fall,
Slow Periwinkle to a crawl,
We'll sit beside our cagéd pool
And let our heated bodies cool.

Or if the mercury drops too far
Make love in our Jacuzzi spa,
While outdoors speakers softly play
To keep raccoons and mice away.

At end of season when the flow
Of sunshine tourists is at a low,
We'll dine in luxury and have fun
With Kiwanis coupons two for one.

Now come, upon the causeway ride,
And watch the pelicans dip and glide
Along the drawbridge that connects
At the same time that it protects

Our magic isle from overuse
By those who nature would abuse,
Who'd hang anhingas out to dry,
Let wounded animals bleed and die,

Let beach erode and wetland drain
'Til just exotic plants remain —
Yet Sanibel has stemmed their tide
And kept its beauty safe inside!

So come, let's live in laid-back style
Upon our nature perfect isle,
A paradise where green dreams dwell —
Retire with me to Sanibel.

Joseph Pacheco is a retired New York City school superintendent and is consultant to the State of Florida. He resumed writing poetry after moving to Sanibel after a creative drought of 50 years.

Dream Homes

"Wouldn't it be nice to live here?" Over a million visitors come to the islands each year. And many, if not most, ask themselves that question. But hardly anyone can pull it off. There's no big city nearby, and few high-paying jobs. And real estate is costly. The median island home price in 2001 was $454,432.

Old Florida revisited. This modern Sanibel home features traditional local architecture

Only a few thousand people make the islands their home. They run local shops or service businesses, or work long-distance out of their homes. Some are retired. Most "residents" are seasonal. They stay on the islands six months and a day, just enough to make them legal residents of Florida, which has no state income tax.

And yes, real estate prices are high. "Many couples ask me 'What can we get for $200,000?'" says Bob Radigan, a local agent. "The answer is maybe a vacant lot." He's exaggerating, but just barely. A two-bedroom duplex sold in March, 2001, for $180,000. A four-bedroom home with a pool was listed for $279,000. But those were worn-down, older buildings. A nice place with modern amenities is tough to find at a good price.

The island market. Other parts of Southwest Florida continue to attract retirees, but Sanibel and Captiva are actually getting younger. The recent economic boom has brought more buyers to the islands in their thirties, forties and fifties. Still in the key earning years of their careers, they are buying a second home here now to get a toehold on the islands, often with plans to move here permanently when they retire.

Fewer island homes are being sold as rental investments. To discourage gentrification, Sanibel doesn't allow homes to be rented out for less than a month, and rental rates have not kept pace with property appreciation. Captiva allows weekly rentals, but homes under $1 million are becoming scarce there.

Older, two-bedroom homes start at about $275,000, if you can find one. Most real-estate ads show houses in the $400,000 to $600,000 range. Homes on the bay and Gulf start in the millions. The few lots left are demanding premium prices, too. A picturesque lot on Sanibel's Dinkin's Bayou, off San Carlos Bay, was $250,000 in 2000. In 2001 it was listed again — at $450,000.

Architectural styles. When we first looked for a home on Sanibel, the agent told us she had a CBS house to show us. Whose could it be, we wondered. We heard Dan Rather had a place here, was that it? Well it turns out "CBS" means "Concrete Block and Stucco," the style of home commonly built in south Florida during the 1960s and 1970s. The outside walls are built out of concrete blocks, covered over with stucco.

There are many CBS homes on Sanibel. Most are ground level and often have a tile roof. And they're durable, except for the flood threat, since they're made out of concrete.

Laws passed in 1979 require all new homes to have their first floor at least 12 feet above sea level. Since the islands average only 4 feet above sea level, most new homes are elevated, set up on wood or concrete pilings. It's an unusual sight for a first-time visitor.

The first wave of piling homes was built in the 1980s. They had exposed wood pilings underneath, vertical board-and-batten wood siding, and asphalt shingles. Later, homes enclosed the piling area, creating an above-ground "basement."

The Old Florida architectural style made a comeback in the 1990s. Built with white or gray clapboard siding and tin roofs, these homes blend well into the island environment. Some have a Key West influence; others have Victorian frills and trim.

Mediterranean Revival architecture also has a presence here. Originally imported (from the northern U.S.) to the Miami area during that city's 1920s real-estate boom, it's common in Sanibel's Sanctuary development.

Building a custom home. You can build a new 1,600-square-foot house on Sanibel for about $325,000, but it'll take some work. Lots now start at $100,000. Construction costs start at $125 a foot, but the best contractors are busy working at nearly twice that price. And be prepared for a lengthy approval process. Sanibel has a long list of building restrictions and requirements, especially concerning native vegetation and wildlife habitat.

Don't even think of building on Captiva unless you're at the top of the market, over $2 million. Worth magazine named the island the 17th most-expensive U.S. real estate market in 2001, just behind Palm Beach.

An easier way to move into a new home is to buy a builder's display or spec home. A wave of these has been built lately, starting at $500,000.

Buying an existing home. Some decent two-bedroom homes — on unpaved roads, relatively far from the beach — are still available on Sanibel for less than $300,000. But they're getting harder to find. Some fall victim to land appreciation; the lot becomes worth more than the house, so owners sell to newcomers who bring in the bulldozers. Others get enlarged to three- or four-bedroom homes, as the current owners, realizing the asset they have, renovate and expand.

You can determine a home's asking price by how close it is to the beach or the water. The most affordable are inland — not on any water, and not within walking distance to the beach. Homes on water now go for at least $500,000, with prices escalating from canals to bayous to bay and Gulf fronts.

Other factors include closeness to the causeway (many folks like being close to the airport) and low maintenance requirements. Homes in gated areas cost more, as do those with private or community pools or views of wildlife areas.

Renovating an older home. The lack of land and affordable construction leads many buyers to run-down older homes they can fix up. Many homes ripe for renovation are in good locations, on canals or near the beach. And they can be affordable.

A New York couple bought an older Sanibel home when they moved to the island in 1997. They paid $220,000 for a CBS home that backed up to conservation land and was just a short walk from the beach. They spent another $150,000 on renovations, putting in a new kitchen, enclosing the back porch, adding an office, replacing the floor and windows, and building a pool. The result? A 2,500-square-foot home, loaded with modern amenities, at a relative bargain price of $390,000.

Buying a rental investment. With today's prices, you'll likely need to put down 30 or 40 percent on an island property to make money renting it out. Rentable condos can make sense, but they are the fastest appreciating real estate on the islands.

Restaurants

"Where to eat?" is a tougher question here. Sanibel and Captiva restaurants are mainly mom-and-pop affairs, each with its own take on menu, service, decor and pricing.

To write this chapter we spent a year eating at nearly every restaurant on the islands. We ate out for breakfast, lunch and dinner, and jotted down our experiences and preferences. But note our biases: we're a middle-class couple with a 7-year-old daughter, prefer simple dishes over gourmet food, and don't eat much meat.

Price codes (one to four dollar signs) reflect the average cost of dinner for one adult including entree, beverage and dessert (tax, tip, or alcoholic beverages not included):

$ = under $10	**$$$** = under $35
$$ = under $15	**$$$$** = $35 and up

(Circled numbers, color-coded by price range, refer to locations on the restaurant map on page 259.)

Credit cards are indicated as follows:

V = Visa	D = Discover Card
MC = MasterCard	DC = Diners Club
AE = American Express	

℺ = **Full bar**

Reviews

㉑ Beachview Steakhouse. American ℺ **$$$** *(Lunch and dinner daily; kid's menu; in the Beachview Estates subdivision (no beach view) at 1100 Parview Dr. , Sanibel; 472-4394; V, MC, D; reservations suggested).* Try the swordfish served crusted with sweet onions or the rack of lamb. Our favorite dessert here: the Granny Smith apple crisp.

㉜ The Bean. American **$** *(Breakfast, lunch, dinner daily; no table service; no smoking; kid's menu; next to the J. Howard Wood Theater at 2240 Periwinkle Way, Sanibel; 395-1919; V, MC).* The newly expanded Bean uses fresh, quality ingredients to make the best salads and sandwiches on the islands, all made to order. The coffee (also sold by the pound) is roasted weekly in small batches and made with great care (the machines are cleaned 10 times a day); the juice is fresh-squeezed. Owners Daniel and Monica Dix hand-select their ingredients daily at local markets. Monica's an artist; she painted the tables and ceilings. As a true mom-and-pop place, the Bean has incredible one-day-only specials now and then; two not to miss are the tangerine-orange juice and the green (hot!) salsa. For breakfast, try a Santa Fe or Mr. Bean bagel, or simply a plain one with cream cheese, onions, tomatoes and sprouts. The delicious muffins are often still warm from the oven. For lunch we get the Cuban bagel sandwich, an Italian panino or any of the huge salads. The Bean has a high kid-friendly quotient, with chalk to draw on the sidewalk, chalk in the bathroom for graffiti and lots of ice cream treats. The new screened patio is our favorite spot on the islands to sit and eat. Our insider advice: try the Bean for dinner. Unlike other island restaurants, it never has a wait after dark (many folks don't realize it's open), and the sandwiches and salads are big enough to be dinner entrees. But bring a sweater on chilly nights, as the patio can get cool.

The Bean

Facing page: Katie Gardenia's

Bubble Room Scout

⑱ Bubble Room. American ⵙ **$$$** *(Lunch and dinner daily; kid's menu; 15001 Captiva Dr. at Andy Rosse Ln., Captiva; 472-5558; V, MC, AE, D, DC).* A ramshackle assortment of small crowded rooms, The Bubble Room has notorious lines, but is a must-see Captiva landmark. Old-fashioned Christmas bubble lights hang on the walls here, along with every other old kitschy item you can imagine. A toy train runs under the ceiling; each table is its own glass-topped display. After your meal, explore every room to marvel at the truly wonderful junk. Servers, called Bubble Scouts, dress in Boy Scout uniforms. Entrees include steak, duck and seafood, but the best food here is the bread basket: cream-cheese Bubble Bread and sticky buns. Desserts are huge; two or three people can share one. We recommend the Tiny Bubble sampler: a salad, choice of appetizer and a dessert. To avoid the lines, come here for lunch.

㊾ The Bungalow. American ⵙ **$** *(Lunch and dinner daily; kid's menu; open until 1 a.m.; 975 Rabbit Rd. at Sanibel-Captiva Rd., Sanibel; 395-3502; V, MC).* This working-class sports bar has finger foods, sandwiches, a raw bar and 11 beers on tap. You can order off the menu of the next-door Island House until 10 p.m. Families are welcome; kids like the pool tables, dart boards and games.

㊷ Chadwick's. American ⵙ **$$** *(Lunch and dinner daily; kid's menu; at the entrance to the South Seas Resort, Captiva; 472-7575; V, MC, AE, D, DC).* Traditional reassuring food, such as fried chicken and prime rib, is the rule here. Buffets include all the usual fixings; the salmon is the best. Desserts feature loads of whipped cream.

㉞ Cheeburger Cheeburger. American **$$** *(Lunch and dinner daily; kid's menu; 2413 Periwinkle Way, Sanibel; 472-6111; V, MC, AE, D).* The huge cheeseburgers at this 1950s diner are cooked to order and have lots of cheese. The creamy shakes are served in frosty tins. Expect a long wait at peak hours. This is the original Cheeburger Cheeburger, which is now a Florida chain.

㉙ Chocolate Expressions. Snacks and desserts **$** *(No tables; in Periwinkle Place, 2075 Periwinkle Way, Sanibel; 472-3837; V, MC).* Yummy treats tempt you everywhere you look: hand-dipped chocolate strawberries, imported chocolates and candies, homemade fudge, gelato, Columbo yogurt and 40 flavors of Jelly Belly jellybeans. Our picks here are the Sun Harvest lemonade and the rich and creamy fruit smoothies, the best on the islands.

⑧ Dairy Queen. Fast food **$** *(No indoor seating; 1048 Periwinkle Way, Sanibel; 472-1170).* A friend of ours calls this the best Dairy Queen in the U.S. We want to agree, though the prices are high and the food is, well, Dairy Queen. But coming here at night, sitting outside on a concrete picnic table and licking down a drippy Dilly Bar does have its charms. Adults are free to relax, as the crowds are exclusively island and tourist families — no carloads of teens. Grandfathered in as the only fast-food restaurant allowed on the islands (now *there's* a business), this DQ is prohibited by Sanibel ordinances from modernizing. So it's still just a walk-up, with the same exact architecture it had when it opened in the 1960s. Somebody ought to declare it a landmark.

Sanibel Dairy Queen

⑫ **Dolce Vita.** Northern Italian ℤ **$$$** *(Dinner nightly; 1244 Periwinkle Way, Sanibel; 472-5555; V, MC, AE, D; reservations suggested).* This cosmopolitan restaurant features dueling pianos each night until 12:30. For appetizers try the crispy fritto misto, cassoulet of Sonoma escargot or the shrimp and avocado gazpacho cocktail. Popular entrees include provimi veal chop Milano, mezzi rigatoni con salsicce, a rustic Italian quesadilla and wild boar.

⑲ **The Dunes Restaurant.** American ℤ **$$$** *(Lunch Mon.–Sat., dinner Wed.–Sat.; kid's menu; 949 Sandcastle Rd., Sanibel; 472-3355; V, MC; reservations suggested).* The public is welcome at this relaxed restaurant overlooking a golf course. For lunch try the Oriental chicken salad or grilled Reuben. Tuesday is all-you-can-eat taco dinner night.

❶ **East End Deli.** Takeout **$** *(Breakfast and lunch daily; no tables; near the lighthouse at 359 Periwinkle Way, Sanibel; 472-9622).* Generous with meat and cheese, the East End Deli makes sandwiches that are perfect for beach picnics. Try the turkey sub with cranberry sauce, the Cuban, or the mozzarella with sun-dried tomatoes on Italian bread. Heart-healthy options include Garden Burgers, a Greek salad wrap and Crazy Bob's (a wrap with chicken, grapes, blue cheese and walnuts). For breakfast ask for an egg sandwich or toasted bagel.

㉛ **Gilligan's.** American/Seafood ℤ **$$** *(Lunch and dinner Mon.–Sat., kid's menu; early-bird specials; 2163 Periwinkle Way, Sanibel; 472-0606; V, MC, AE, D; call-ahead seating).* This simple one-room restaurant has the islands' best bargain: a swamp cabbage (hearts of palm) salad with dill dressing for $2.49. The Mexican specials are worth a look; our favorites are fish tacos (grilled fish with taco fixings in a soft shell) and black bean quesadillas. Other picks: the po-boy sandwich and the Steambag (snow crab, shrimp, clams, oysters and mussels served with corn on the cob and fries). If smoke bothers you sit away from the bar.

❺ **Gramma Dot's Seaside Salon.** Seafood **$$** *(Lunch and dinner daily; kid's menu; at the Sanibel Marina, 634 North Yachtsman, Sanibel; 472-8138; V, MC, AE).* This waterside institution attracts a loyal, older clientele. Good bets include the seafood Caesar salad, coconut or bacon-wrapped shrimp, mahi-mahi, mesquite grouper or Maine lobster (reserve one in advance). Homemade chips come with all lunches. For dessert try the key lime pie or homemade chocolate cake.

㉝ **The Green Flash.** Seafood ℤ **$$$** *(Lunch and dinner daily; kid's menu; early-bird specials; 15183 Captiva Dr., Captiva; 472-3337; V, MC, AE, D, DC; reservations suggested).* Otters often play in the water at this waterside restaurant. At lunch try the Green Flash sandwich (turkey or vegetables and cheese on grilled focaccia), or the barbecued shrimp and bacon. Good starters for dinner include oysters Rockefeller and shrimp bisque. For dinner try the garlicky grouper Cafe de Paris or the salmon with dill Bernaise sauce. The restaurant is on the bay side of Captiva, so you won't see a green-flash sunset here. But you get a great view of the tranquil Roosevelt Channel and Pine Island Sound.

㉝ **Greenhouse Grill.** Mediterranean **$$$** *(Lunch and dinner daily; kid's menu; outside patio dining for smokers; 2407 Periwinkle Way, Sanibel; 472-6882; V, MC, AE; reservations suggested).* This island standout had fallen on hard times, but now it's back with new owners and a reinvigorated spirit. Diane Badalich and Chef Carlo DiSomma (from the now-departed Bellini's on Captiva) have lavished their new creation with care and detail. Ask Carlo if he uses fresh ingredients — you'll see the passion in his answer. Highlights include a classic bouillabaisse locals swear by, a 16th-century escargot recipe discovered from the Borgia family, and spicy Shrimp Piccanti. For lunch try one of the huge sandwiches with homemade skin-on fries. Desserts are flown in from Milan. It's worth a trip here just for the bread and its spicy oil dip.

41 Happy Shrimp. Seafood **$$** *(Breakfast, lunch and dinner daily; kid's menu; in the Olde Sanibel Shops at 630-1 Tarpon Bay Rd., Sanibel; 472-6300; V, MC).* The menu at this hideout includes shrimp pastas, skillets, curries and a fine (though messy) traditional barbecue shrimp cooked in beer, honey and spices. Portions can be small, but the recipes are good. Colorful murals line the walls; the center bar has seashells and other sea life embedded in the top. Few people know about this place; there's never a wait.

37 Hungry Heron. American **$$** *(Breakfast, lunch and dinner daily; kid's menu; across from Eckerd at 2330 Palm Ridge Rd., Sanibel; 395-2300; V, MC, AE, D).* Kids love the TVs tuned in to Nickelodeon and the Disney Channel here. The food comes in generous portions; the menu has over 250 selections. Best bets include the baked Beach Bread and roast beef or Hot-to-Trot sandwiches. Monday-night prime rib specials and an all-you-can-eat weekend breakfast buffet during season are good values.

10 Huxters Market and Deli. Takeout **$** *(Breakfast, lunch and dinner daily; no tables; 1203 Periwinkle Way, Sanibel; 472-6988; V, MC, AE, D).* This family-owned store features fried chicken, salads, generous sandwiches and homemade desserts. Huxters says it sells the coldest beer on the island, at 28° F.

48 Island House. American/Seafood **Y $$$** *(Lunch and dinner daily; kid's menu; 975 Rabbit Rd. at Sanibel-Captiva Rd., Sanibel; 472-8311; V, MC; reservations suggested).* Co-owned by Sanibel City Councilman Marty Harrity, this friendly mid-island restaurant has great service and big comfortable booths. Lunch specials start at $5.99. Dinner options include a nice pecan-crusted or crunchy grouper and gingerbread pork. Try the Chinese cheese stix for an appetizer. Kids will love the Gingerbread House, an

Island Inn dining room

ooey-gooey dessert of homemade gingerbread, vanilla ice cream, chocolate and caramel sauce and whipped cream.

47 Island Inn. American **$$$** *(Breakfast and (seasonal) dinner daily; nonsmoking; 3111 West Gulf Dr., Sanibel; 472-1561; V, MC, AE, D, DC; reservations suggested).* Those staying here always have first priority on tables, but squeezing in breakfast or (formal) dinner at this Sanibel landmark will be one of your more unique experiences. This is true Old Florida; not much has changed for decades. Breakfasts are a set $10, dinners a set $24 (including appetizer and dessert); the menu changes daily. All the food is homemade. Call first; don't expect to make it in during holidays and the peak season.

18 Island Pizza and Pasta. Italian **$$** *(Lunch and dinner daily; kid's menu; delivery; 1619 Periwinkle Way, Sanibel; 472-1581; V, MC).* The tight booths of this 1970's-era hangout remind you how much we've all fattened up since then, but it's rarely crowded and the food is fine. The garlic bread is warm and soft. The best bet is the pasta, specifically the spaghetti carbonara, penne San Remo or angel hair Caprese.

11 Jacaranda. Seafood **Y $$$** *(Dinner nightly; early-bird discounts; 1223 Periwinkle Way, Sanibel; 472-1771; V, MC, AE, D, DC; reservations suggested).* A loyal, affluent crowd keeps the Jacaranda packed. This civilized restaurant boasts a lovely outdoor patio and attentive service. Start with Sancap Shrimp, four jumbos sautéed with spiced rum, coconut cream and Florida orange juice. Top entrees include Florida snapper en papillote and chicken Giovanni. Tuna is a local favorite. After dinner there's live music and dancing in the Patio Lounge.

㊸ Jean Paul's French Corner. French $$$$ *(Dinner Mon.-Sat. Dec.-April; next to the Sanibel Post Office at 708 Tarpon Bay Rd., Sanibel; 472-1493; V, MC; reservations suggested).* Intimate and romantic, Jean Paul's offers French cuisine prepared with care and imagination. The menu includes salmon, duck, scallops, veal, daily specials and a great wine list. Try the Panaché de Canard aux Cassis de Dijon (duck leg confit and magret with a delicate black-currant sauce), Saint-Jacques Antiboise (sea scallops with saffron sauce) or Saumon Frais (Norwegian salmon with dill sauce). Kids love the "big lemon with ice cream in it."

㉕ Jerry's Restaurant. American $ *(Breakfast, lunch and dinner daily; kid's menu; inside Jerry's Supermarket, 1700 Periwinkle Way, Sanibel; 472-9300; V, MC).* The no-nonsense waitresses have accents straight from Minnesota (Jerry's home base) at this 1970's-era supermarket diner, where the food and value are tough to beat. The BLTs have big tomato slices; the coconut cream pie is delicious. Kids love the spaghetti and the Bodacious Brownie. The fresh salad bar is a plus, too. For breakfast try the French toast. Giant booths overlook a tropical courtyard.

㊴ Johnny's Pizza. Takeout $ *(Lunch and dinner daily; free delivery; no tables; 2496 Palm Ridge Rd., Sanibel; 472-1023).* Our picks at this Chicago-style pizzeria are the garlicky white pizza and the Grandma Rose Pizza (with a romano and parmigiana base covered with crushed plum tomatoes and topped with cheddar and mozzarella).

㉘ Katie Gardenia's. New American/Seafood ✗ $$$ *(Lunch and dinner daily; kid's menu; in the Forever Green Shops, 2055 Periwinkle Way, Sanibel; 472-1242; V, MC; reservations suggested).* The Orchid Room patio here is one of the prettiest dining spots on the islands. But inside the setting's still relaxing and interesting, with island memorabilia and hundreds of variations of mermaids filling every nook and cranny. Katie founded Captiva's Bubble Room; her

whimsy and imagination continue here in a modern, romantic setting. The sophisticated menu has unusual dishes, with lots of fresh seafood. Start with crispy blue-crab fritters or chili con queso. For lunch try the Cuban sandwich with a mojo dipping sauce or the lemon rosemary rotisserie chicken on a sourdough baguette. For dinner we get the flaming (!) cedar-planked salmon or the Caribbean seafood stew in a roasted tomato, spinach, coconut and lemongrass broth. Wines are selected especially for each dish. Leave room for dessert: the Katie Kake has four layers of chiffon cake, fresh whipped cream, homemade praline candy and slivers of toasted almonds; the Orange Crunch layer cake is made with Grand Marnier and local oranges, topped with orange cream-cheese frosting and citrus slices, and finished with marmalade.

㊶ Keylime Bistro. New American $$$ ✗ ✓ *(Lunch and dinner daily; live outdoor entertainment; 11509 Andy Rosse Ln., Captiva; 395-4000; V, MC, AE, D, DC; reservations suggested for larger parties).* Want gourmet food in a shorts and T-shirt atmosphere? This is your place. Where else can you eat crab cakes with a key-lime aoli (spicy and garlicky), or stuffed shrimp in a lobster sauce (rich and creamy), while you sit outside listening to reggae music under a tiki hut? Keylime Bistro makes its own sushi; the tuna tataki and inside-out tuna roll are standouts. The creative salads are first-rate. The key lime cheesecake is the best dessert on the islands. Made in layers with ladyfingers soaked in key lime liqueur, it's light as air, but huge; enough for three people. On weekends stop in late; they serve appetizers, salads and desserts until midnight.

⑰ La Vigna. Italian $$$ *(Dinner daily; no smoking; 1625 Periwinkle Way, Sanibel; 472-5453; V, MC; call-ahead seating).* This white-tablecloth restaurant has first-rate pasta and the best gourmet pizza (cheesy and garlicky) on the islands, served in a comfortable, relaxed atmosphere. Locals recommend the vegetable pizza.

⑦ Lazy Flamingo. American/Seafood $$ *(Lunch and dinner daily; kid's menu; just west of the Sanibel Causeway at 1036 Periwinkle*

Way, Sanibel; 472-6939; also at Blind Pass on 6520 Pine Ave., Sanibel; 472-5353; V, MC, AE, D). Built as a combination fish house and sports bar, the large Lazy Flamingo on Periwinkle is also the island's best family restaurant. In fact everyone, regardless of age or class, eats here. Expect a long wait anytime from 5 p.m. to 9 p.m. during season, and even a short delay any other time of the year. The fish is straight off the boat and cleaned in the back. Straightforwardly prepared (no fancy sauces needed here, thank you), this is the best seafood on the islands. If we had to choose our last meal, it would be served in a big booth here and involve blackened grouper, pasta marinara and a few bottles of the frigid beer they keep iced behind the bar. Locals rave about the teriyaki hot wings. Other great choices are the peel-and-eat shrimp, clam chowder, hamburgers and grilled chicken sandwich, and any nightly special, especially the tuna and homemade crab cakes. We substitute a small Caesar salad for the fries that come with every meal. The waitresses here are the best: friendly and incredibly attentive. Kids get crayons and a coloring sheet. Island children rate the Buffalo wings as their favorite restaurant food. The original Lazy Flamingo 🔟, a tiny 7-booth bar at the west end of Sanibel, has the same menu, though there's no nonsmoking area. It fills quickly after sunset. While you're there try your hand at the ring game, though island kids may shame you by hooking the ring before you do. In our survey of islanders, 59 percent of adults (and 67 percent of kids) name the Periwinkle Way Lazy Flamingo as one of their three favorite restaurants.

②︎ Lighthouse Cafe. American **$$** *(Breakfast, lunch and (seasonal) dinner daily; kid's menu; no smoking; in the Seahorse Shops, 362 Periwinkle Way, Sanibel; 472-0303; V, MC; call-ahead seating)*. Sanibel native Mike Billheimer's island landmark serves all three meals, but there's a reason the sign out front says "The World's Best Breakfast." Bring your heartiest appetite for rich Eggs Benedict, whole wheat granola hotcakes, or the Lighthouse Special, an omelet with turkey, broccoli, mushrooms and cream cheese topped with homemade hollandaise sauce.

Ask for a booth; they're big and worth a wait. The fast service will get you in and out in no time, but in season get here before 8 a.m. to avoid a long wait. For lunch try a fish sandwich, hamburger or salad. Reasonably-priced dinners are served from mid-December through Easter, and the casual fare includes fresh fish and sandwiches.

㊿ Mad Hatter. New American **$$$$** *(Seasonal lunch and dinner daily; no smoking; at the western tip of Sanibel at Blind Pass, 6467 Sanibel-Captiva Rd., Sanibel; 472-0033; V, MC, AE, D; reservations suggested)*. Don't let the tiny barn-like exterior fool you. Like Alice through the looking glass, you walk into the Mad Hatter and up into a magical pink, white and green dining room overlooking the Gulf for one of the better gourmet experiences in all of Florida. An ideal place for a romantic dinner with a bottle of wine, it has a subtle Alice in Wonderland theme that doesn't get in the way. The Mad Hatter has an innovative flair with California, Southwest and Southern cuisines, using recipes that combine strong and mild flavors. The menu changes monthly, with no dish repeated, but you'll always get a wonderful basket of crusty breads (our favorite is the sun-dried tomato). We loved the Seacakes appetizer (lump blue crab meat, crawfish, red bell pepper and onion, served with honey mustard sauce and mango salsa), and the jumbo sea scallops entree (achote brushed and pan-seared with a marvelous citrus vanilla bean Chardonnay sauce, capers and green onions served with spinach ricotta pasta shells and grilled veggies). For dessert try the Crème Brûlée (custard topped with caramelized sugar), made with real vanilla bean instead of extract so there's no hint of bitterness. There's no kid's menu, though the kitchen keeps some basic pasta and chicken on hand.

㊶ Mama Rosa's Pizzeria. Italian **$** *(Lunch and dinner daily; kid's pizza; no table service; delivery; across from the South Seas Resort in Chadwick's Square, Captiva; 472-7672; V, MC)*. Come to this small pizzeria for its sinfully decadent Beach Bread appetizer: garlic bread layered with ricotta cheese and tomato, topped with melted Monterey jack

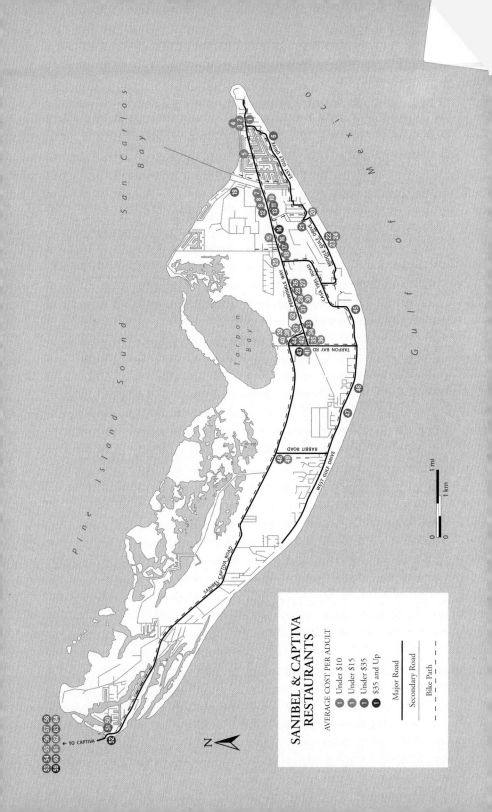

and cheddar cheese. The pizza's good too; hand-tossed with toppings that include sun-dried tomatoes and artichoke hearts. Any pizza can be prepared "white," without tomato sauce. The menu also has calzones, subs and salads.

9 Matzaluna. Italian ϒ $$ *(Dinner daily; kid's menu; 1200 Periwinkle Way, Sanibel; 472-1998; V, MC, AE, D, DC; call-ahead seating).* Decorated with food stock and Italian flags, this upbeat two-story restaurant fills up early during the season. The crusty bread with garlic dipping sauce is a hit; the freshly grated Romano cheese is sharp and delicious. Pasta is cooked perfectly al dente, not chewy or hard; our favorite is the ravioli. The fresh seafood specials are a good bet. Cooked in a wood-burning oven, the delicious pizzas here have a great crust. The superb item is the rich crab bisque, available only on Fridays and Saturdays. Island kids recommend the ravioli, hamburgers and squid. Butcher-paper tablecloths and complimentary crayons inspire would-be artists; Matzaluna has drawings hung on the walls from its customers, including a signed sketch from Stephen King. For dessert we usually get the simplest of all: a bowl of rich, creamy vanilla ice cream. But the key lime pie is also outstanding.

13 McT's Shrimp House. Seafood ϒ $$$ *(Dinner daily; kid's menu; early-bird discounts; next to the 7-Eleven at 1523 Periwinkle Way, Sanibel; 472-3161; V, MC, AE, D, DC; call-ahead seating).* The all-you-can-eat steamed shrimp platter is especially popular at this casual Old Florida restaurant. For dessert, mud pie is an infamous, heady concoction of Oreo cookies, chocolate fudge, coffee ice cream, heavenly hash ice cream and whipped cream. The first 100 persons in the door after 4 p.m. qualify for early-bird specials. In the back, McT's Tavern has a pinball machine, a huge sports TV, appetizers and light dinners. Notice the bar ceiling, built of upside-down Australian pine roots.

38 Mona Lisa's Pizzeria and Deli. Italian $ *(Lunch and dinner daily; free delivery; 2440 Palm Ridge Rd., Sanibel; 472-0212; V, MC).* Set in a small strip mall next to Goodwill, this sit-down pizza parlor has a bright but comfortable no-frills atmosphere and good service. Order honest pizza by huge slices or by the pie. Unusual toppings include meatballs, ricotta cheese, breaded eggplant and spinach. The deli subs have lots of meat and cheese; also available are pastas, salads, Italian tiramisu and, of course, key lime pie. The $1.95-pizza slice is a true bargain, as are the $3.95 lunch specials.

20 Morgan's Forest. American ϒ $$ *(Breakfast and dinner daily; kid's menu; at the Holiday Inn, 1231 Middle Gulf Dr. at Donax St., Sanibel; 472-4100; V, MC, AE, D, DC).* Young children get a kick out of this dense "forest" of plastic plants. Stuffed animals, ground-level fog and thunder and lightning effects add to the kitschy fun. The most popular dinner entrees are the Tropical Forest Grouper En Papillote (baked in parchment with shrimp, spinach, artichoke hearts and garlic butter), the Caribbean-flavored Cajun Shrimp & Chicken Belize Linguine; the Matambre (Argentine Rolanade) tenderloin stuffed with shrimp, peppers, cheese and scallions; and the tasty Amazon Crab Cakes. Bacon and eggs are the best selections on the breakfast buffet.

57 Mucky Duck. American/British/Seafood $$ *(Lunch and dinner Mon.–Sat.; kid's menu; no smoking; 11546 Andy Rosse Ln., Captiva; 472-3434; V, MC, AE, D, DC).* Victor Mayeron and his family always have a kind greeting for you at their waterfront restaurant, which combines a warm, friendly atmosphere with cheeky British humor (tip: don't complain if you don't sit by the window). The fare is American and British comfort food and fresh seafood. At lunch, go for the charred and tasty Pub burger — a great value. At dinner it's all good; islanders especially like the crab cakes. Don't miss the fluffy and creamy key lime pie — it's made with fresh whipped cream daily by Mrs. Mayeron, and we think the best on the islands (though Victor often parades around the room with free forkfuls of other desserts at dinner time). With its low ceilings, bare-wood floors and corny signs and posters, this little hideaway on the Gulf could pass for a cozy country inn. In fact it used to be the Gulf View Inn, back in the 1930s. A recent real-estate boom has bought up much of Andy Rosse Lane, but the Mayerons march on, refusing to sell. Don't miss the patio and beach (ideal for sunsets), where there's live music every night but Sunday.

㊱ Nick's Place. Takeout **$** *(Snacks daily except Sunday; no tables; in the Bailey's Shopping Center at 2477 Periwinkle Way, Sanibel; 472-0770).* This hole-in-the-wall (literally) offers Columbo yogurt, frozen custards, sundaes, milk shakes, banana splits and smoothies. If you don't have a sweet tooth, try a cappuccino, espresso or latte.

㉓ Noopie's. Japanese **$$** *(Dinner nightly; kid's menu; at the Sundial Resort, 1451 Middle Gulf Dr., Sanibel; 395-6014; V, MC, AE, D, DC; reservations required).* The authentic Teppanyaki Japanese dining here features seafood and steaks prepared on a table-grill in front of you. Noopie's Special Dinner is a good choice if you'd like to try a variety of dishes; a sushi appetizer, Japanese soup, Oriental salad, shrimp complement, two entree choices, stir-fry vegetables, fried rice, bean sprouts, pea pods and Noopie's sundae for $28.95. The servers prepare the sizzling fare with style and flourish.

㊾ Old Captiva House. Continental/Seafood ♈ **$$$** *(Breakfast and dinner daily; kid's menu; no smoking; at 'Tween Waters Inn, 15951 Captiva Dr., Captiva; 472-5161; V, MC, AE, D; reservations suggested).* This romantic restaurant takes you back to a more gracious era. The very definition of Old Florida style; everything is green and white, with hardwood floors, soft indirect lighting, shaded candles on the tables, French doors and windows all around, and live piano. Famed political cartoonist and environmentalist J.N."Ding" Darling and his wife ate here routinely; his original drawings hang on the walls. For appetizers try the sweet Captiva bisque, a creamy blend of shrimp, brandy and herbs; or Oysters Florentine, among the best we've had. The rich Chicken Wellington is also good, but too much for one person: two huge chicken breasts stuffed with an artichoke Florentine, baked in puff pastry and finished with a gorgonzola cream sauce. The bread is the best on the islands, soft and crusty with butter as light as air. The excellent key lime pie is more solid than many versions, topped with dribbles of four fruit sauces. The child's crunchy grouper dinner is huge: four big planks of fish, a square of cheesy potatoes au gratin, and fresh grilled squash and other veggies.

㉚ Periwinkle Place Bistro. American **$$** *(Breakfast, lunch and dinner daily; kid's menu; early-bird discounts; in the Periwinkle Place shopping center, 2075 Periwinkle Way, Sanibel; 472-2525; V, MC; reservations suggested).* Formerly the Sanibel Island Chowder Co., this comfortable cafe still makes a soothing chowder. But this is also the best (almost only) place on the island for Mexican food, with fine quesadillas and burritos (our choice is the vegetarian burrito, a big, hearty blend of black beans, rice, sautéed vegetables, melted cheddar cheese, wrapped in a flour tortilla and served with pico de gallo). For dinner try the Ropa Vieja, a traditional Cuban meal of spicy shredded beef simmered with onions and peppers.

④ Pinocchio's. Ice cream **$** *(No tables, benches outside; near the lighthouse at 362 Periwinkle Way, Sanibel; 472-6566).* The claim-to-fame at this tiny shop is the delicious, not-too-sweet Sanibel Krunch ice cream, with nuts and coconut. All the ice cream is homemade; special touches include using whole nuts, not crushed pieces, in the pistachio. Kids can't resist the Dirt Cup; soft-serve chocolate-vanilla ice cream, crushed Oreos and chocolate sprinkles, with gummi worms peeking through. Other temptations: chocolate-dipped strawberries and handmade fudge. The coffee bar serves espresso, cappuccino, Cuban coffee and iced coffee as well as the regular cup of joe, all ground and brewed to order.

㉖ Pippin's. American ♈ **$$$** *(Dinner nightly; early-bird specials; kid's menu; in Tahitian Gardens, 1975 Periwinkle Way, Sanibel; 395-2676; V, MC, AE, D, DC; reservations suggested).* This newly reopened old reliable features a traditional menu of steaks, prime rib, ribs, seafood and a big salad bar. The

R.C. Otter's

homemade Death by Chocolate Cake is an indulgence; desserts are served with multiple forks. Kids love the tropical fish tanks.

⑥ Portofino. Northern Italian ¥ **$$$** *(Breakfast and dinner daily; kid's menu; at the Sanibel Inn, 937 East Gulf Dr., Sanibel; 472-0494; V, MC, AE, D, DC; reservations suggested).* Many of the recipes here have been handed down through generations. A good appetizer is the unusual Ostriche Alla Sambuca (oysters on the half-shell baked with sautéed spinach, anisette, bacon and mozzarella). For an entree try the Salmone Al Formaggio, a fresh salmon filet filled with herbed ricotta, broiled with Chardonnay and topped with pesto butter. The bar has live piano entertainment.

㉕ R.C. Otter's. American **$$** *(Breakfast, lunch and dinner daily; kid's menu; 11506 Andy Rosse Ln., Captiva; 395-1142; V, MC, AE, D; call-ahead seating).* Lots of island feel here: A funky painted building, live laid-back music and front-porch and patio dining. Created by the original owners of the Hungry Heron, Rob and Cathy DeGennaro (hence the "R.C."), Otter's has another huge menu with items to please anyone. We recommend starting with the cold strawberry bisque, or the black beans and rice topped with salsa, sweet onions and sour cream. Good sandwiches include the Cuban, barbecue pork and grouper Reuben. For bigger appetites try the crab-and-black-bean burrito or the blackened scallops.

㊻ Riviera. Mediterranean/Northern Italian ¥ **$$$** *(Dinner nightly; kid's menu; ½ mile west of Tarpon Bay Rd. at 2761 W. Gulf Dr., Sanibel; 472-1141; V, MC, AE; reservations suggested).* The dark, romantic setting, with live piano, is ideal for the menu of interesting dishes. Appetizers include baked oysters, made with spinach, ouzo, cream and Swiss cheese. The surprising pistou soup has fresh vegetables, cranberry beans, tubettini pasta and spinach pesto. Exotic entrees follow, such as Lamb Tajine, stewed in terracotta, ginger, coriander, caramelized onions, prunes and couscous. Or try the Chicken Riviera, stuffed with Montrachet cheese, spinach, green peppercorns and basmati rice. Kids love the Rock Candy dessert.

③ Rosie's Island Market & Deli. Takeout **$** *(Breakfast, lunch and dinner daily; no tables; delivery; next to the Lighthouse Cafe at 362 Periwinkle Way, Sanibel; 472-6656; V, MC, D).* Ken and Leslie Barker have taken this small market and deli and turned it into a true restaurant-to-go. Breakfasts feature fresh-baked bakery items, pancakes and omelets; lunch offerings include seafood baskets, deli sandwiches (big enough for two, with soft buns), hamburgers, hot dogs and pizza. Dinners-to-go include comfort food (meatloaf, mashed potatoes, fried chicken) and Tex-Mex. Specials vary daily. For dessert try a Pelican Poop, a ball of ice cream covered in chocolate; flavors include cappuccino, bing-cherry cheesecake, toasted coconut and peanut butter cream. Rosie's offers free delivery anywhere on Sanibel between 11 a.m. and 2 p.m.

㉗ Sanibel Cafe. American **$$** *(Breakfast, lunch and dinner daily; kid's menu; in the Tahitian Gardens shopping center, 2007 Periwinkle Way, Sanibel; 472-5323; V, MC; call-ahead seating.)* A cozy cafe where each table is a shadow box of shells, the Sanibel Cafe excels at breakfast. Best are the banana pancakes and the coconut French toast (both so tasty they don't need syrup). Also first-rate are the ham and Havarti cheese omelet and the black raisin bread with cream cheese and homemade red raspberry jam. The lunch menu features a different homemade soup daily, plus the island-favorite Rusty Pelican sandwich. Dinner choices include traditional meals like meat loaf, grilled ham and cheese and hamburgers. You can also order from the lunch menu for dinner.

⑭ Sanibel Steakhouse. American ⅄ $$$$ *(Dinner nightly; 1473 Periwinkle Way, Sanibel; 472-5700; V, MC, AE, D, DC; reservations suggested).* For the carnivore in you, try the hand-cut USDA Prime Grade (i.e., best two percent of meat available) steaks here. Seafood and chicken entrees are also available. Try the peanut butter pie for dessert.

⑮ Schnapper's Hots. Takeout $ *(Breakfast, lunch and dinner daily; kid's menu; no tables; 1528 Periwinkle Way, Sanibel; 472-8686; also next to R.C. Otter's at 11508 Andy Rosse Ln., 472-3647).* These people love, honor and cherish hot dogs. Exuberantly friendly, Schnapper's char-grills hot dogs, burgers and sausages so good they pop when you bite into them (our favorites are the Polish sausage and bratwurst). The fries are prepared fresh and cooked to order. There's officially no place to sit (government rules), but we always cheat and eat on the benches. All kids get a free small toy from the goodie box. The Captiva Schnapper's ㊽ opened in the fall of 2001.

㉟ The Seafood Factory. American ⅄ $$ *(Lunch and dinner daily; kid's menu; in the Bailey's Shopping Center, 2499 Periwinkle Way, Sanibel; 472-2323; V, MC, AE, D, DC; call-ahead seating).* You can order seafood, steaks or pasta, but seafood is the specialty. The large menu has over 100 items, including seafood platters, Carolina she-crab soup, herb-crusted grouper and Tuna Caliente, blackened tuna filet with black beans, salsa, sour cream, scallions, and black olives served over rice with tortilla chips. A delicious signature dish is the Greek shrimp feta pasta. A separate menu with 50 more items is available in the Factory Lounge, which is opened from 11:30 to late at night.

㉔ The Shoppe. Takeout $ *(Lunch and dinner daily; no tables; at the Sundial Resort, 1451 Middle Gulf Dr., Sanibel; 472-4151 ext. 3840).* This small but well-stocked store has deli specials such as fresh-baked pizza, hot entrees, hearty sandwiches, salads, Danish pastries, bagels and gourmet coffees. This is a great spot to stop and get a bite while cruising Sanibel on a bike.

㊵ Subway. Takeout $ *(Lunch and dinner daily; no tables; 2496 Palm Ridge Rd., Sanibel; 472-1155).* One of the few franchises on the islands, Subway offers basic, healthy sandwiches and salads at a good price. Breads are baked fresh four times a day. In typical island fashion, hours here are sporadic: the attendant takes a break a couple times a day, and closes up for a half-hour or so.

Where islanders eat

We surveyed 586 full-time Sanibel and Captiva residents — young singles to retirees, working class to über-rich, meat lovers to vegetarians — about their restaurant preferences. Their top 10 restaurants, in order, are:

1. Lazy Flamingo*
2. Trader's
3. Island House
4. Timbers
5. Lighthouse Cafe
6. La Vigna
7. Matzaluna
8. Hungry Heron
9. Riviera
10. Sanibel Steakhouse

We also surveyed island kids (ages 5 to 18). Their favorites:

1. Lazy Flamingo*
2. Hungry Heron
3. Island House
4. Matzaluna
5. Subway
6. Timbers
7. Dairy Queen
8. Cheeburger Cheeburger
9. Schnapper's Hots
10. Bungalow

*Lazy Flamingo is far and away the top place to eat among locals, getting four times the votes of any other restaurant, from adults and kids alike.

Grill. Seafood/Steakhouse ☨ $$$ lunch and dinner daily; at Blind Pass at 6536 Pine Ave., Sanibel; 472-2333; V, MC, AE, D; reservations suggested). An antique copper ceiling, open-air kitchen and pecky-cypress paneling gives this tiny place a relaxed charm. Every table faces Turner Beach. This is Lazy Flamingo-founder Larry Thompson's "serious" restaurant, with a chef preparing gourmet-quality meals and an impressive wine selection. Any nightly special will be delicious, but our favorite meal is straight off the menu: the filet mignon, a melt-in-your-mouth steak wrapped in prosciutto and topped with a rich Bearnaise sauce, served on a bed of sautéed mushrooms. Another good choice is the chicken piccata, pan-seared boneless chicken breast topped with roasted red peppers, capers and key lime butter. For appetizers try the Maryland crab cakes or broiled baby brie (with toasted almonds, honey dijon and fresh fruit). The terrific black bean soup is stocked with sausage. Lunches are just as good — quality sandwiches and that killer black bean soup — and a better value. But the real secret here is the breakfast. The pancakes and Eggs Benedict are the best on the islands, as they come from the same gourmet kitchen. There's never a crowd at breakfast except on Saturday mornings, when the island bike club treks up here.

60 Sunshine Cafe. American/Steakhouse $$$ (Lunch and dinner daily; kid's menu; Captiva Village Square, Captiva; 472-6200; V, MC; reservations suggested). An open-air kitchen adds to the cozy feel at this intimate California-inspired eatery. The menu changes frequently and offers a handful of fresh selections for dinner. Lunches include Thai salad, hamburgers and wood-grilled pork loin. Dinners include sesame-crusted yellowfin tuna salad, Colorado lamb chops, and nightly fresh-fish specials.

45 Thistle Lodge. American/Seafood ☨ $$$ (Lunch and dinner daily; kid's menu; early-bird specials; at the Casa Ybel Resort, 2255 West Gulf Dr., Sanibel; 472-9200; V, MC, AE, D, DC; reservations suggested). A great view complete with a beachfront sunset awaits you at the Casa Ybel restaurant. Try the island blue crab cake for lunch. The most popular dinner entrees are the Snook Island Paella, Bronzed Red Snapper with Mango Papaya Sauce and Jamaican Jerk Pork.

42 Timbers/Sanibel Grill. Seafood and Steakhouse/Sports Bar ☨ $$ (Dinner daily; kid's menu; across from the post office at 703 Tarpon Bay Rd., Sanibel; 472-3128; V, MC, AE, D, DC; call-ahead seating). This light and airy casual restaurant offers no surprises: just seafood and steaks, cooked to order, in a nice, comfortable atmosphere. There's a long list of fresh catches and specials daily; the coconut cake is a good dessert. Next door the Sanibel Grill sports bar (open until midnight) offers good pizza and sandwiches (the best is the char-grilled chicken sandwich with mango chutney sauce), "chicken lips," and a tasty spinach dip.

16 Trader's. New American/Seafood $$$ ✓ (Lunch and dinner daily; kid's menu; no smoking; 1551 Periwinkle Way, Sanibel; 472-7242; also at 5050 Captiva Dr., Captiva; 395-9494 (operating as "Viva, a Trader's Bistro") ; V, MC, AE, D; reservations suggested). The dining room has cool, clean air at Trader's, and the food is fresh, interesting and unusual. No need to look at the menu; the specials are what you want, especially the Cajun scallops. Other specials can be macadamia grouper with a Thai peanut sauce and fresh mango salsa, blackened mahi-mahi with roasted corn salsa and a lime remoulade, and horseradish-crusted swordfish with a roasted red pepper sauce. If they are serving a crisp for dessert, get it hot with vanilla ice cream. After your meal explore the 8,000-square-foot store filled with unusual home furnishings, accessories, books and gifts. On Captiva, **Viva 63** has the same menu in a modern setting.

44 Twilight Cafe. Fusion/New American/ Vegetarian $$$ (Dinner Mon.–Sat.; kid's menu; no smoking; behind Tower Gallery at 751 Tarpon Bay Rd., Sanibel; 472-8818; V, MC; reservations suggested.) Local artists have hand-carved the walls and ceilings at this eclectic, whimsical but romantic cafe, the best gourmet-level restaurant on the islands. For starters try the homemade corn and crab chowder with roasted red pepper and cilantro ribbons, or grilled portobella and tomato bruschetta with gorgonzola and a petite arugula salad. Imaginative dishes include grilled tuna on Chinese greens and mushrooms with mango basmatti and a kiwi-honey glaze, grilled steak with sweet mashed potatoes and a roasted green apple glaze, and a grilled veal chop in a caramelized onion and sherry reduction with corn and crawfish mashed potatoes. Islanders recommend the grilled pork chops and the Beef Medallions of Diane with grilled broccoli. There are novel vegetarian entrees, too. An open-air kitchen and wood-fired grill add a homey touch.

㊾ Village Cafe. French **$$$$** *(Dinner nightly; 14970 Captiva Dr., Captiva; 472-1956; V, MC, D; reservations suggested).* With curved white stucco walls and an open kitchen, this intimate restaurant serves good French food. An extensive wine list includes some unusual choices such as a Peju Provence Chardonnay from California. Good choices include any special, the sautéed fresh Foie Gras, the Duck Maple Bourbon and the Truffled Free-Range Young Roaster.

㉒ Windows on the Water. Florasian/Seafood ⵗ **$$$** *(Breakfast, lunch and dinner daily; kid's menu; at the Sundial Beach Resort, 1451 Middle Gulf Dr., Sanibel; 395-6014; V, MC, AE, D; reservations suggested.)* This long, narrow room has a wall of windows looking out over the Sundial's large swimming pool and the Gulf. Stepped levels ensure there's no bad seat in this calm Asian-influenced restaurant. The "Florasian" menu ("Florida favorites with a Pacific Rim flavor") features a good Sanibel Dim Sum (seasoned shrimp and chicken wrapped in steamed wonton, served with a sweet soy and garlic sauce). Other top choices include Marathon Mahi-Mahi (sesame crusted with a light honey garlic sauce) and Chicken Cayo Costa (with sun-dried tomatoes, mozzarella and wild mushroom ragout). Even if you're not staying at the Sundial, we recommend coming here for breakfast. The $10 buffet is a fine eye-opener, and the morning view is terrific.

Our Best Bets

Breakfast
1. Banana pancakes, Sanibel Cafe
2. Eggs Benedict, Sunset Grill
3. Santa Fe bagel, The Bean
4. Coconut French toast, Sanibel Cafe
5. Lighthouse Special, Lighthouse Cafe

Lunch
1. Blackened grouper sandwich, Lazy Flamingo (above)
2. Tenderloin sandwich, Sunset Grill
3. Italian panino, The Bean
4. Pub burger, Mucky Duck
5. Barbecue pork sandwich with black beans and rice, R.C. Otter's

Dinner
1. Blackened grouper platter, Lazy Flamingo
2. Crab cake platter, Lazy Flamingo
3. Filet mignon, Sunset Grill
4. Jumbo sea scallops in vanilla bean Chardonnay sauce, Mad Hatter
5. Any special, Trader's

Dessert
1. Keylime cheesecake, Keylime Bistro
2. Crème Brûlée, Mad Hatter
3. Key lime pie, Mucky Duck
4. Katie Kake, Katie Gardenia's
5. Sanibel Krunch ice cream, Pinocchio's

♦ Kid's picks*
1. Deviled crab special, Lazy Flamingo
2. Dirt Cup, Pinocchio's
3. Vanilla malt, Cheeburger Cheeburger
4. Ravioli, Matzaluna
5. Black bean quesadillas, Gilligan's

* Choices of Micaela Neal, age 7

Shopping

Shopping on the islands is special. And that's both good and bad.

Competing for the fickle tourist dollar, each store tries to be unique. That means you'll have no trouble finding unusual tropical shirts and dresses, solid-gold sand dollar earrings and hand-painted furniture — all high quality. But need something mundane — a camcorder battery, a pair of plain jeans — and you'll have quite a hunt.

That gripe aside, the stores and galleries offer all sorts of hidden delights and one-of-a-kind keepsakes, and the quality is first-rate. Nearly all are mom-and-pop shops; the owners' passion for their hand-picked merchandise is contagious.

Art galleries

Aboriginals: Art of the First Person *(in the Village, 2340 Periwinkle Way, Sanibel; 395-2200)* features museum-quality tribal art from Africa, Aboriginal Australia, native America and the Arctic people of Canada and Alaska, authentically sourced from tribal culture. Selections include ancestral carvings, pottery, baskets, textiles, jewelry and ceremonial masks.

You'll find one of Southwest Florida's largest selections of museum-quality antique and period jewelry at the charming **Albert Meadow Antiques** *(across from the Bubble Room at 15000 Captiva Dr., Captiva; 472-8442)*. The impressive variety of diamond, gold and platinum estate pieces complements original Maxfield Parrish lithographs and collections of Art Nouveau glass from Tiffany, Steuben, Daum and Galle.

Contemporary art is the focus of the **Black Orchid Gallery** *(in the Timbers Center, 705 Tarpon Bay Rd., Sanibel; 472-8784)*. Styles range from the vibrant colors of Hessam and Sabzi to the subtle tones of Picasso. Also available are dozens of Limoges boxes, exotic dolls, whimsical frogs, the delicate hand-blown glass of Novano, and the etched crystal of Sanibel's Luc Century.

Operating in an old island cottage, the **Hirdie Girdie Gallery** *(2490 Library Way at Tarpon Bay Rd., Sanibel; 395-0027)* offers Sanibel-themed oils, watercolors, pastels, acrylics, baskets, tiles and wood carvings.

Once part of the Tower Gallery, **Island Framing and Gallery** *(next to the Hungry Heron restaurant, 2330 Palm Ridge Rd., Sanibel; 472-1118)* does first-class frame work. Denise McEnroe also stocks artwork, including paintings and painted furniture.

The funky **Island Style** gift shop *(in Periwinkle Place, 2075 Periwinkle Way, Sanibel; 472-6657; and across from the South Seas Resort at 210 Chadwick's Square, Captiva; 472-4343)* features expensive handmade furniture, glass, pottery, jewelry, sculpture, and wall art. Most have a tropical folk-art theme that reflects the relaxed style and joy of the islands. The fanciful wooden tables, in particular, are unique pieces.

Jan's Antiques and Art *(next to the Bean, 2242 Periwinkle Way, Sanibel; 395-0200)* has pottery, blown glass, art jewelry, antique furniture and marble, stone and wood flooring. Paintings include Realism, Impressionism and modern art selections.

Huge metal sculptures rest in the trees outside of **Jungle Drums** *(11532 Andy Rosse Ln., Captiva; 395-2266)*. Inside, the collection of wildlife and environmental art includes animals in bronze, wood, ceramics and glass, plus jewelry, framed art, prints, books, furniture and music. Over 250 artists are featured.

Owner Ruth Beyersdorf showcases unusual and beautiful art at **Kelly's Cocoons** *(14830 Captiva Dr., Captiva, 472-8383; 2496 Palm Ridge Rd., Sanibel; 395-0422)*: Kelly's butterfly art, Diane Mannion's pottery and hand-painted floor cloths, John Mannion's painted furniture, and blown glass by Kellmis Fernandez.

Invitational and juried shows feature works from a variety of Florida artists at the **Phillips Gallery** *(inside the Barrier Island*

Facing page: Decorative basket at Tower Gallery, by Gisela Damandl

Furniture at Seaweed Gallery

Group for the Arts (BIG Arts), 900 Dunlop Rd., Sanibel; 395-0900). The exhibits change monthly, and all the art is for sale. The gallery and art receptions are free. Don't miss the adjacent outdoor sculpture garden. BIG Arts is the islands' nonprofit cultural arts center.

Sanibel Art and Frame (2460 Palm Ridge Rd., Sanibel; 395-1350) offers a large selection of Sanibel and Captiva prints, photographs and posters. Services include next-day custom framing, shipping and free island delivery and hanging.

The oldest gallery on the islands, the **Sanibel Gallery** (in the Heart of the Island Shops, 1628 Periwinkle Way, Sanibel; 472-3307) represents over 170 local artists and craftspeople. Oils, watercolors, clay, glass, wood, metal, paper and the annual Sanibel Christmas ornament reflect the tropical island environment.

The Seaweed Gallery (in the Forever Green shops, 2055 Periwinkle Way, Sanibel; also next door to the Keylime Bistro at 11509 Andy Rosse Ln., Captiva; both stores: 472-2585) has an eclectic, charming collection of island pottery, stained glass, oil, watercolors, acrylics, jewelry, pen and ink, art dolls and

painted furniture. Mermaids are a common theme. A Captiva branch, opened in fall 2001, features outdoor exhibitions.

Teresa Kostrubala's Art Studio (in the Village, 2340 Periwinkle Way, Sanibel; 472-8006) features tropical scenes and contemporary abstracts created with brilliant colors and metallic paints. Choose from prints, greeting cards, posters and originals. You can watch Teresa work from the window.

Tin Can Alley (2480 Library Way, Sanibel; 472-2902) is a work of art itself. Artist Bryce McNamara has sectioned off a corner of his funky home to sell hammered, punched, and sculpted pieces of recycled tin, which he has transformed into fish, suns and other objects. The luminaries produce magical patterns of light. Lamps, recycled furniture and tropical folk art are also for sale.

A Touch of Sanibel Pottery (1544 Periwinkle Way, Sanibel; 472-4330) hand-crafts functional and decorative stoneware and porcelain, and colorful raku. Local and nationally-recognized potters are featured, so there's a wide choice of colors and designs.

Filling both stories of a brightly painted historic beach house, **Tower Gallery** (751 Tarpon Bay Rd., Sanibel; 472-4557) is a co-op of 20 local artists. A well-chosen assortment of quality paintings, prints, raku ceramics, baskets, painted furniture, photography, stained glass and handcrafted jewelry is constantly changing. A member artist usually sits behind the counter. Our favorite items here are Marty Stokes' fish prints (made from real fish) and Teri Causey's funky (and affordable) tables and cabinets.

A combination gallery and gift shop, the **Tree House** (in the Olde Sanibel shops, 630 Tarpon Bay Rd., Sanibel; 472-1850) features local paintings, photography and pottery, including Shelligrams, Hilda Kaihlanen's Island Primitives and Cat's Meow Village miniature collectibles.

Beach paraphernalia

Beach Stuff (14900 Captiva Dr., Captiva; 472-3544) has the best floats on the islands; many hang from the overhang outside the store (they'll even blow up your floats and rafts for you after you buy them). Inside you'll find swimsuits, skim boards, towels, beach shoes and sandals, kites, beach toys, masks, hats, sunglasses, sunscreen and a big selection of T-shirts. Gifts include the Sanibel lighthouse ornament and hand-painted "Soul Fish" by Bonnie Murray, and those classic Florida kitsch souvenirs.

Sanibel Surf Shop *(in Jerry's Shopping Center, 1700 Periwinkle Way at Casa Ybel Rd., Sanibel; 472-8185)* has a variety of moderate-priced beach supplies, plus surfing equipment and a good selection of books, games, toys and puzzles.

The largest selection of beach stuff is at **Winds** *(2353 Periwinkle Way, Sanibel; 395-0091)*. The store sells inexpensive and moderate-priced swimsuits and T-shirts for all ages, as well as boogie boards, masks, fins, beach towels, tote bags, sunscreen, floats and a few seashells.

Bookstores

The three island bookstores overflow in Sanibel spirit. Each has its own personality, run by an islander who loves to read:

Primarily a paperback exchange, the **Island Book Nook** *(2330 Palm Ridge Rd., Sanibel; 472-6777)* has over 10,000 new and used paperbacks. Joan Simonds puts the books she recommends on a table in front. She stocks a comprehensive list of local books and field guides, and *rents* hardcover bestsellers for $5 a week.

MacIntosh Book Shop *(2365 Periwinkle Way, Sanibel; 472-1447)* has been Sanibel's main bookstore for decades. Owner Jim Dowling has a friendly, knowledgeable staff. Jim will search the world (i.e., the Internet) to find the book you want. "We find the most remote books on earth!" he tells us. The shelves include local fiction, field guides and children's books.

Hollie Smith's **Sanibel Island Bookshop** *(across from Jerry's Supermarket, 1711 Periwinkle Way, Sanibel; 472-5223)* reflects her laid-back sensibility: her dog may greet you when you walk in. Her shop is filled with every definition of a "beach-read" — local authors, field guides, Oprah's choices. The children's section is especially appealing. Hollie's greeting cards and stationery are among the most interesting on the islands. She also stocks an assortment of unusual gifts, including lightstone crystals and miniature hand-blown vases.

Four Sanibel attractions also sell books. The **Bailey-Matthews Shell Museum** gift shop *(3075 Sanibel-Captiva Rd., Sanibel; 395-2233)* has great shelling books, plus gifts and decorative items with a shell theme. The children's section has some terrific items. The bookstore at the **J.N. "Ding" Darling National Wildlife Refuge** *(inside the visitors center, 1 Wildlife Drive, Sanibel; 472-1100)* has a great selection of nature and field guides, nature videos, wildlife-themed gifts and toys, and refuge T-shirts and hats. The **Sanibel-Captiva Conservation Foundation** *(3333 Sanibel-Captiva Rd., Sanibel; 472-2329)* has an extensive gift shop, with hard-to-find nature and field guides. Finally, the **Sanibel Historical Village and Museum** *(950 Dunlop Rd., Sanibel; 472-4648)* sells books on Sanibel history (some out of print) inside (appropriately) the old Bailey's General Store.

Clothing

You'll find a huge selection of swimwear and swimming accessories at the **Beach House** *(in Periwinkle Place, 2075 Periwinkle Way, Sanibel; 472-2676; and across from the South Seas Resort in Chadwick's Square, Captiva; 472-4665)*. Over 5,000 suits for men, women and children pack the store. The women's selection is a Godsend — coverage options from tiny bikinis to conservative, yet still flattering, choices; sizes up to 30; mastectomy, long-torso and D-cup styles; and you can buy tops and bottoms separately. Accessories include cover-ups, sunglasses and Nike goggles.

C. Turtles and Company *(next to the Bean, 2242 Periwinkle Way, Sanibel; 472-1115)* has bright and colorful men's and women's casual clothes. Brands include Jams World, Tori Richards and Tommy Bahama.

Island Artists

Ikki Matsumoto is an island institution. His playful, abstract wildlife prints hang in businesses and homes throughout Sanibel and Captiva. The Tokyo native moved to the United States in 1955; he later studied under artist Charles Harper. Today he creates soft watercolors and is a fixture at island galleries and art fairs.

Marty Stokes and his family use the ancient Japanese art form of Gyotaku ("fish impressions") to create unusual and beautiful paintings. Originally developed as a way of measuring the exact size of a caught game fish, Gyotaku obtains a printed impression of a fish by painting it with block inks and pressing absorbent paper over it. The technique records the natural beauty of the fish, while still giving the artist creative freedom. Marty's version of the process combines soft unryu, mulberry bark, chiri, and other handmade rice papers with water-based inks. He uses seahorses and other sea life to create underwater scenes. His work is sold at the Tower Gallery and other island shops.

Bryce McNamara uses discarded tin and other metals to create whimsical and functional objects. His lamps, lanterns and luminaries feature punched designs that cast intricate light patterns. Fanciful fish and animals are popular for children's rooms. See Bryce's work at his gallery and workshop, Tin Can Alley.

Luc Century uses a photo-stenciling process to etch wildlife images onto glass sculpture, paperweights, glassware and wall art. It's available at the refuge gift shop and many island galleries. The Vietnam Memorial used his technique to etch its names onto stone.

Goz Gosselin creates elaborate "flower" arrangements out of seashells. The former flower-shop owner also makes mirrors, table lamps and floor lamps. Look for his work at the Black Orchid Gallery and the Sanibel Shell Fair. A large arrangement is on display at the Sanibel Public Library.

Teri Causey makes funky hand-painted furniture and wall hangings, featuring mermaids and other tropical motifs. It's available at the Tower Gallery and Seaweed Gallery.

We should also point out that internationally renowned artist **Robert Rauschenberg** lives on the islands, and participates occasionally in local exhibitions.

Left: Dragonfly by Bryce McNamara. **Right:** Storage cabinet by Teri Causey. **Top:** Gyotaku painting by Marty Stokes

Caribbean Coast *(in Periwinkle Place, 2075 Periwinkle Way, Sanibel; 472-2993)* carries upscale tropical clothing for men and women, with a good selection of Tommy Bahama apparel and shoes.

Casual Attitude *(in Tahitian Gardens, 2001 Periwinkle Way, Sanibel; 472-0088)* has classic outdoor and travel clothing for men and women, plus jewelry. Brands include Royal Robbins, Hook & Tackle, Tori Richards and Brighton.

No wonder women love **Chico's** *(in Periwinkle Place, 2075 Periwinkle Way, Sanibel; 472-0202; at Palm Ridge Place, 2330 Palm Ridge Dr., Sanibel; 472-3773; and across from the South Seas Resort in Chadwick's Square, Captiva; 472-4426)*. Here you can be a size 2, even though you're really a 12! The store's unique sizes run 0-3, matching the traditional range of 6-16. The lightweight clothes are versatile and fashionable. The Sanibel Chico's are the original stores of this now-national chain.

Comfort by Design *(in the Lime Tree Center, 1640 Periwinkle Way, Sanibel; 395-0666)* offers quality shoe and sandal brands such as Birkenstock, Mephisto and Ecco. Inserts are available; shipping is free.

Manager Sally Kopplow stocks the **Cricket Shop** *(in the Anchor Point Shopping Center, 1633 Periwinkle Way, Sanibel; 395-2277)* with thoughtful choices. Swimsuits include father/son and mother/daughter designs, and all children's sizes. There's a good selection of contemporary sportswear and classic dresses. We like the crushable straw hats and "no-headache" visors.

Dockside Quality Clothing and Shoes *(in Periwinkle Place, 2075 Periwinkle Way, Sanibel; 472-9098)* carries upscale, outdoor men's clothing, with a few items for women. Most pieces have a fish or outdoor theme. Brands include Timberland (shoes and clothing), Reyn Spooner (apparel, including FSU and UF tropical shirts), Cutter & Buck, Hook & Tackle, Newport Blue, Ex Officio, Guy Harvey (shirts and hats), Teva (sandals), Freestyle (watches) and Costa del Mar (sunglasses). It's a great-looking store, too.

The classic cut of the apparel at **Eileen Fisher** *(in the Forever Green shops, 2055 Periwinkle Way, Sanibel; 472-4655)* flatters all shapes and sizes. Sophisticated women's clothing and accessories use neutral colors and lightweight, comfortable fabrics.

You'll find island-casual footwear for the whole family at **Footloose** *(in Jerry's Shopping Center, 1700 Periwinkle Way at Casa Ybel Rd., Sanibel; 472-4717; and across from the South Seas Resort in Chadwick's Square, Captiva, 472-1707)*. Brands include Sperry, NAOT, Minnetonka (moccasins), Teva (sandals), Speedo (beach shoes) and Costa del Mar (sunglasses).

Fresh Produce *(in Periwinkle Place, 2075 Periwinkle Way, Sanibel; 395-1839)* sure has the island spirit, though the small chain is based in Boulder, Colo. The brightly colored, all-cotton clothing is extremely soft and comfortable. Many pieces have funky animal or nature designs. Matching accessories include socks, ponytail bands and caps.

H$_2$0 Outfitters *(in Periwinkle Place, 2075 Periwinkle Way, Sanibel; 472-8890; and across from the South Seas Resort in Chadwick's Square, Captiva; 472-7507)* stocks outdoor and travel apparel for men and women, including T-shirts, shorts, jeans and sandals. Quality brands include Tommy Bahama, Weekender, Nautica, Ex Officio, Guy Harvey and Royal Robbins. Shoe brands include NAOT, Sperry and Teva. They also carry Costa del Mar sunglasses.

Her Sports Closet *(in Periwinkle Place, 2075 Periwinkle Way, Sanibel; 472-4206)* features casual island sportswear, in sizes 2 through 18 and with many mother/daughter coordinates. Brands include Lilly Pulitzer. Check out the unusual belts and purses.

Children's beachwear is available at next-door neighbors **Island Beach Co.** and **Island Beach, Too** *(14820 Captiva Dr., Captiva; 472-3272)*. The boutiques sell swimsuits, coverups, T-shirts and shorts for all ages. There's a surf shop inside, too.

The upscale apparel at **Island Pursuit** *(in Periwinkle Place, 2075 Periwinkle Way,*

Sanibel; 472-4600) includes soft island suits for men. Brands include Helen Kaminski, Tommy Bahama, Brighton and Axis. Shoe lines include Sperry, Cole Haan and Bragamo. Purses, belts and other accessories round out the selection.

Glamorous, arty, Moroccan, Indonesian... many adjectives describe **I-Spy** *(2340 Periwinkle Way, at The Village Specialty Shops; 472-2221)*. Owner Pat St. Cyr hand-picks women's clothing and jewelry that have a definite exotic style. Unusual accessories include delicate shawls, 1920s-era Bakelite jewelry and mah-jongg tile bracelets.

Brightly colored cotton separates for women and kids stock the **Keylime Clothing Co.** *(across the street from the Lighthouse Cafe, 359 Periwinkle Way, Sanibel; 395-1870)*. You'll also find funky Sanibel T-shirts and hats, books, gifts, cookbooks, sandals, beach toys and picture frames.

The upscale children's clothes at **Lads & Lassies** *(in Periwinkle Place, 2075 Periwinkle Way, Sanibel; 472-1180)* include many unusual pieces, some handmade. Accessories, including fun sunglasses, and toys are tucked in there, too.

The **Lion's Paw** *(near the causeway at 1025 Periwinkle Way, Sanibel; 472-0909)* features elegant silk and tencel dresses and separates. Casual and formal fashions offer style and sophistication.

Jackets are a specialty at **Lookin Good** *(in the Olde Sanibel shops, 630 Tarpon Bay Rd., Sanibel; 472-6888)*. Unusual choices include hand-painted canvas and decorated denim. Also note the soft, shimmery ribbon shirts from Korea.

Classy, dressy **Maggie Elliott** *(in Periwinkle Place, 2075 Periwinkle Way, Sanibel; 472-2230)* offers clothes that combine a big-city elegance with a vacation spirit. The boutique carries the complete line of Brighton accessories, jewelry and shoes.

Mango Bay *(in Jerry's Shopping Center, 1700 Periwinkle Way at Casa Ybel Rd., Sanibel; 472-6678)* carries clothing and beach items for the whole family, includ-

T-shirts at Mango Bay

ing quality T-shirts and hats, swimsuits, shorts, sunglasses, sandals and beach shoes. Both tiny and conservative swimsuits are available for women.

Memories of Sanibel *(in the Olde Sanibel shops, 630 Tarpon Bay Rd., Sanibel; 395-1410; and next to the Bean at 2240 Periwinkle Way, Sanibel; 395-0990)* carries shorts, hats and accessories, as well as a good selection of quality Sanibel T-shirts, including many embroidered styles.

Beautiful, high-end children's apparel for lucky children 0 to 14 fills **Nanny's** *(in the Village, 2340 Periwinkle Way, Sanibel; 472-0304)*. The store has exquisite handmade items, plus shoes, jewelry, books, toys, even dress-up clothes.

We were shocked when we walked in **Pandora's Closet** *(near the lighthouse at 455 Periwinkle Way, Sanibel; 395-2400)*. From the outside you'd never know it, but inside are some of the most gorgeous, unusual outfits you'll find anywhere. From ultra-formal evening dresses you could wear to the Oscars to wonderful Victorian hats and colorful beaded handbags, owner Karen Leonardi has exquisite taste. Casual wear and high-end children's clothing round out her inventory.

Paradise of Sanibel *(in Periwinkle Place, 2075 Periwinkle Way, Sanibel; 472-3020)* has one of the best selections of island T-shirts. You'll also find big tropical shirts for men, soft cotton sweaters for women, plus clothes for kids and babies. Other items include Beanie Babies, jewelry, frames, chimes, toys and gifts. Parrot Heads will feel at home — the store has Jimmy Buffett music on the stereo constantly, sells Buffett's lines of T-shirts (sizes medium to 4X), beach towels and note cards, as well as nearly every Buffett album (even some hard-to-find vintage discs from the early '70s). The shop is worth a look just for the decor: Owners Roger Digby and his sister Marilyn Bodimer have made the walls and floor a matching landscape; flamingos wrap around you in the dressing room.

The only plus shop on the islands, **PlusPerfect** (in the Forever Green shops, 2055 Periwinkle Way, Sanibel; 472-8110) stocks stylish, lightweight tropical attire from Emme, Karen Kane, Amanda Gray, CMC Cotton, Kedem Sasson and Flax. Separates let you mix and match. Don't overlook the nice lingerie, accessories and jewelry selected by owner Judy Dearborn.

The intimate **Shirley Allen** boutique (in the Olde Sanibel shops, 630 Tarpon Bay Rd., Sanibel; 472-4544) has one-of-a-kind, high-priced pieces. The fashionable clothes are elegant and comfortable. Lovely original paintings adorn the walls.

The **Sporty Seahorse** (362 Periwinkle Way, Sanibel; 472-1858) is the island's department store, a 1960's-era throwback. Much bigger than it looks, it has a large selection of apparel and accessories for men, women and children, including over 2,000 swimsuits. Nothing trendy here, just good quality and price. Look for bargains — clearance merchandise is tucked in throughout the store.

One of the largest apparel stores on the island, **Stanley & Livingston's Island Outfitters** (in the Village, 2340 Periwinkle Way, Sanibel; 472-8485) specializes in lightweight classic shorts and shirts for men and women. The selection also includes hats, sandals, native jewelry, accessories and T-shirts from brands such as Ex Officio, Royal Robbins, Brighton and Picante.

The **Tahitian Surf Shop** (in Tahitian Gardens, 2015 Periwinkle Way, Sanibel; 472-3431) has apparel for men, women and children, including Tommy Bahama clothing. The swimsuit collection is small, but includes daring styles not found elsewhere. Brands include Ray Ban (sunglasses), Freestyle (watches) and Reef (beach shoes).

At the **T-Shirt Place of Sanibel** (in Periwinkle Place, 2075 Periwinkle Way, Sanibel; 472-2392) the name says it all: a good group of funky T-shirts for men, women and kids. They also carry shorts, hats, soft cotton separates, beach totes and fun jewelry.

When our relatives come to the island and want a souvenir T-shirt or cap, the first place we mention is the **West Wind Surf Shop** (in Periwinkle Place, 2075 Periwinkle Way, Sanibel; 472-3490). The organized store also sells sunglasses, skim boards, water toys, hats, books and jewelry. Brands carried include Billabong, Tommy Bahama, Oakley (sunglasses), Freestyle (watches) and Reef and Quiksilver (sandals).

Why Knot (in the Village, 2340 Periwinkle Way, Sanibel; 472-3003) sells soft, lightweight womenswear (brands include Eileen Fisher) made of natural fibers such as cotton and flax. The washable linens are especially popular. The solid-color T-shirts are worth a look, too. Accessories include Jeanine Payer sterling silver jewelry.

Convenience stores

Both **7-Eleven** stores (1521 Periwinkle Way, Sanibel; 472-9197; and at intersection of Tarpon Bay Rd., 2460 Periwinkle Way, Sanibel; 472-8696) are clean, friendly spots with good selections of ice cream bars, plus to-go bakery products and hot dogs. The Tarpon Bay Road store has pay-at-the-pump gas at off-island prices.

It's 10 miles between Sanibel's Tarpon Bay Road and Captiva's Andy Rosse Lane, and the only store in between is **Hal's Grocery** (6406 Sanibel-Captiva Rd., Sanibel; 472-5227). The lack of competition shows — a half gallon of milk costs $3.50. But Hal's has its merits: a good variety of ice cream bars and treats, and it's right by Turner Beach.

The newly-remodeled **Hess Express** (at Tarpon Bay Rd., 2499 Palm Ridge Rd., Sanibel;

472-2198) is corporate-showcase clean and has pay-at-the-pump gas. No island ambience here, but Hess is customer-focused: The pumps are under a large canopy (vital during a summer downpour), prices are the same as off-island locations and the tire air is free. Inside there are convenience items as well as to-go sandwiches, hot dogs and ice cream.

Huxters Market and Deli *(near the causeway at 1201 Periwinkle Way, Sanibel; 472-2151)* is the first store as you come on the island; the last one before you leave. Here you'll find all the standard convenience items plus beach supplies, video rentals, liquor, cigars, paperbacks, newspapers and an old-fashioned deli with sandwiches, fried chicken, salads and desserts.

You'll be surprised how much stuff is packed inside Captiva's tiny **Island Store** *(11500 Andy Rosse Ln., Captiva; 472-2374)*. Like an old general store, there's a little of everything here — a few hard-boiled eggs over here, some frozen bait over there. Groceries, rental videos, fine gifts, apparel and to-go sandwiches fill every available space. Don't overlook the delicious homemade Jensen's Ice Cream. Next door is an even tinier liquor and cigar store.

Rosie's Island Market & Deli *(in Old Town Sanibel, 362 Periwinkle Way, Sanibel; 472-6656)* serves Sanibel's east end, offering groceries, convenience items, Old Town Sanibel posters and postcards, baked goodies and ice cream treats. There's good take-out food here, too.

Hidden inside the lobby of the Sundial resort, **The Shoppe** *(1451 Middle Gulf Dr., Sanibel; 472-4151)* is a pretty good small convenience store — handy for bikers riding down East Gulf Drive. Selections include international cheeses and freshly-baked cookies.

Gifts

Arundel's Hallmark Shoppe *(in the Heart of the Island shops, 1626 Periwinkle Way, Sanibel; 472-8317)* is a one-hour photo shop, a FedEx shipping center, and the island's office supply store. Oh yeah, it's a Hallmark shop, too, with the usual fine-quality greeting cards, ornaments, gifts and photo albums.

Bandanna's at Sundial *(inside the Sundial Resort, 1451 Middle Gulf Dr., Sanibel; 472-4151)* features casual swimwear, tennis apparel, cover-ups, casual dresses, shorts, accessories, shoes and tennis equipment.

We've lived on Sanibel for years but never stopped in **Caloosa Canvas** *(1616 Periwinkle Way, in the Heart of the Island shops, Sanibel; 472-2218)* until researching this book. We figured it sold awnings or tents. But "canvas" here refers to its stock of bags, beach chairs, hammocks, umbrellas and pillows. They also have some fun metal garden accessories and painted bird houses.

Island-themed stained glass and lots of local paintings and prints fill **Captiva's Finest** *(across from the South Seas Resort at 110 Chadwick's Square, Captiva; 472-8222)*. There's sculpture, too, as well as replica Sanibel and Captiva street signs. This is the sister store of Sanybel's Finest.

Carmen Lombardo has turned his home into the **Confused Chameleon** *(11528 Andy Rosse Ln., Captiva, 472-0560)*. The 1950s Captiva cottage is now a whimsical gift shop, run by Carmen and his partner, Constantine Stratos. Standout items include retro balancing toys with mechanical movement, votive candle holder birds, spiritual cards by Susan Miller and delicate mermaid watercolors. Many of the pieces are from local artists.

Owner Daniel Moore-Thompson describes **End Result** *(in Periwinkle Place, 2075 Periwinkle Way, Sanibel; 395-3333)* as "an upscale Pier One without the furniture and with jewelry." Standouts include delicate hanging glass balls, sterling-silver jewelry, Indian paper lamps, beaded barrettes, ceramics, wind chimes and funky Christmas ornaments. Every time we're in here we feel like adding a rec room to our house, just so we could fill it with these things.

We keep expecting to see Martha Stewart at the **Golden Pear** *(2407 Periwinkle Way,*

Islander Trading Post

Sanibel; 472-5681); it's her kind of store. Graciousness and class fill the air here, along with heavenly potpourri. Upscale housewares and gifts include unique dishes, ceramics, candles, stationery, flowers, lotions and hats.

Il Crocodile (at Palm Ridge Place, 2330 Palm Ridge Dr., Sanibel; 472-9166) is worth a wander as you're waiting for a table at the Hungry Heron. The eclectic collection has home furnishings, hostess gifts, toys, Vera Bradley handbags, original handcrafted skirts by Ruth Comes, creative ethnic and trouves jewelry by Lannie Cunningham and the Fitz and Floyd Gallery of dinnerware, gifts and Christmas accessories.

You'll go back in time at the **Islander Trading Post** (1446 Periwinkle Way, Sanibel; 395-0888). Old signs, radios, kitchen items, curios and seemingly every other old thing is for sale here, at modern prices (why oh why didn't we hang onto that stuff in Grandma's attic!). There are some new items, too: Sanibel-Captiva blankets, and great candles.

Island Gifts and Shells (1609 Periwinkle Way, Sanibel; 472-4318) is an old-fashioned gift shop, with many shells, shell bottles and shell-themed gifts, books, postcards, towels, knickknacks and Florida T-shirts.

The owner of **Jonna's of Sanibel** (at Palm Ridge Place, 2330 Palm Ridge Dr., Sanibel; 472-2302) laughs as she describes her pieces as "wearable art," but that's what it is: Lovely hand-painted clothing and jewelry. Jonna's is the exclusive shop on the island for a line of jewelry called Lunch at

the Ritz. Each pin, necklace or pair of earrings is one-of-a-kind. The White House ordered some pieces a few years ago (perhaps intern gifts?); other famous customers include Janet Jackson. Jonna also carries some wonderful sculpture, wildly painted and decorated with faces and flowers. And don't miss her collander-and-teaspoon lamps and mobiles.

Beautiful glass sculptures are the standouts at **The Mole Hole of Sanibel** (in the historic Cooper house in the Olde Sanibel shops, 630 Tarpon Bay Rd., Sanibel; 472-2767). Choose from dolphins, elephants, lovebirds, even multicolored sun catchers. The jewel-toned perfume bottles are worth a look, too, as are the hollow witches balls to hang in the window (they ward off "evil spirits").

Quirky figurines, statues, mobiles, ornaments, jewelry, glass, ceramics, kid's books and teddy bears fill every nook and cranny of **Pandora's Box** (in Periwinkle Place, 2075 Periwinkle Way, Sanibel; 472-6263). Greeting cards are tucked upstairs. One-of-a-kind pieces abound, many homemade.

Close your eyes and you'll think you're in heaven at the rejuvenating **Petals Boutique & Bicycles** (next to Cheeburger Cheeburger, 2427 Periwinkle Way, Sanibel; 472-4546); it smells great here. Eclectic merchandise includes English teas, soft Grecian scarves and French aromatherapy paper. The kid's section has clothes, swimsuits, stuffed animals and unique Brazilian animal masks. Petals is also a rental agent for Billy's Rentals (Petals owner Sally Kirkland is Billy's wife); you can rent a bike from here and have it delivered to your hotel or condo.

The **Sandpiper of Sanibel** (next to the Hungry Heron restaurant at 2330 Palm Ridge Rd., Sanibel; 472-4645) features sterling silver jewelry and accessories for the home, wall hangings, bird carvings, baskets, pottery, gifts, cards and lamps.

Wall-size maps and charts are just one of the unique items at **Sanybel's Finest** (in Jerry's Shopping Center, 1700 Periwinkle Way at Casa Ybel Rd., Sanibel; 472-6776). Local watercolor artist Barney Baller is well represented, as is wood carver Rod Becklund. Other items include mirrors, fountains and shell bottles and lamps.

Kassia Strauss offers unique furnishings and accessories at her **Secret Garden** *(in the Village, 2340 Periwinkle Way, Sanibel; 395-0600).* Tucked in the back of the Village, the store's absolutely worth finding for its unusual pieces, many from local artists, including hand-painted furniture, custom floral arrangements and trees, salt and pepper shakers, floor cloths and pillows.

A top-of-the-line apparel and gift boutique, the **South Seas Shoppe** *(South Seas Resort, Captiva; 472-1994)* features shirts with the South Seas logo or that say "Captiva." The artistic items include whimsical studio glass from Anna Ornberg, unusual hand-painted bed linens from Susan Sargent, hand-painted furniture and jewelry boxes for children. They also do custom apparel work for groups, and welcome baskets.

Unique primitive furniture and artifacts from Mexico, Indonesia and Morocco surround the diners at **Trader's Store and Cafe** *(1551 Periwinkle Way, Sanibel; 472-7242).* The large store also includes scented candles, clothing, books, toys and aromatherapy bath essentials.

Tuttles Sea Horse Shell Shop *(next to the Sporty Seahorse and the Lighthouse Cafe at 362 Periwinkle Way, Sanibel; 472-0707)* is packed with kitschy Florida souvenirs. But there's more serious stuff, too, such as local artist "Apple Annie" Rothwell's hand-painted wooden benches and Sanibel's Dave Terlap's handmade wooden seahorse tables and crab benches. Rounding out the goods are beach paraphernalia, T-shirts, sweats, hot sauces, Florida's Fabulous books, shells, sunglasses, 14-karat gold jewelry and some nice Sanibel Christmas ornaments.

Wilford & Lee *(in Tahitian Gardens, 2019 Periwinkle Way, Sanibel; 395-9295)* recently doubled its space, and manager JoAnn Perry has filled it with many interesting home accents and gifts. Choose from wood and metal wildlife sculptures, hostess trays, tile-topped tables, shell bottles, framed island-themed art and those omnipresent Ty Beanie Babies. UPS shipping is available.

Groceries

There's no Jimmy Buffett music playing at **Bailey's General Store** *(in Bailey's Shopping Center, Periwinkle Way at Tarpon Bay Rd.; Sanibel; 472-1516).* It's way too authentic for that. Founded in 1899 by Frank Bailey and run today by his sons, Bailey's is the heart and soul of the islands. From the street it looks like a simple combination grocery-and-hardware store. But Bailey's history as a true general store has made it a catch-all for whatever islanders need. Here they'll change your watch battery, send your telegram, fill your prescription, propane tank or fishing line, ship your fruit, or repair your vacuum cleaner. They ship UPS packages now, too, and develop film. Don't miss the to-die-for homemade fudge (at the coffee bar at the front of the store), made daily with cream and butter. The bakery makes a dandy Hurricane sandwich and has excellent key lime pie (it should — the Baileys used to farm key limes). The hardware side of the store sells fishing supplies, live bait (shrimp), fishing licenses and a few appliances, toys and gifts. Every Sanibellian seems to have different reasons to love Bailey's. Ours: they have handles on their paper bags, and they give kids free cookies. Bailey's charm is most visible during Baileyfest, the free-food carnival it throws for islanders each fall, and during a tropical storm, when it's one of the few places that stays open. And, yes, the legendary Bailey brothers are still here, running it all. The thin one is Sam, the jolly one is Francis.

CW's Market & Deli *(South Seas Resort, Captiva; 472-5111)* serves South Seas guests. It has a little of everything, including an extensive deli, liquor, and video rentals. Fresh-ground coffee is sold by the cup or the bag.

Macrobiotic, organic and natural foods fill the shelves of **Island Health Foods** *(1640 Periwinkle Way, Sanibel; 472-3666).* They've got takeout, too: fresh carrot juice, protein drinks, smoothies and organic snacks. Also here: bulk food such as nuts and grains, organic free-range eggs, soy milk, vitamin and mineral supplements, aromatherapy supplies and homeopathic herbs.

Besides typical supermarket goods, **Jerry's Foods** *(in Jerry's Shopping Center, 1700 Periwinkle Way at Casa Ybel Rd., Sanibel; 472-9300)* has a good salad bar, hot and cold takeout, a varied cheese selection, unique juices and a post office on-site. But we like it for the "underground" parking (underneath the elevated building), out of the sun (and thunderstorms) and unique service: After you check out, a conveyor belt sends your groceries down to the parking area, to a drive-through lane. You pull up, clerks load your trunk, and off you go!

Hardware

Bailey's True Value Hardware *(in Bailey's Shopping Center, Periwinkle Way at Tarpon Bay Rd.; Sanibel; 472-1516)* is part of Bailey's General Store. See Bailey's listing under Groceries, above.

Forever Green Ace Hardware *(in the Forever Green shops, 2025 Periwinkle Way, Sanibel; 472-5354)* has more than paint and hardware. There's a good selection of imported pottery and an extensive lawn and garden shop. Unusual Mosquito Magnets kill mosquitoes and no-see-ums without chemicals (a portion of profits go to SCCF). And if you need a snook mailbox (the door is the mouth), Sanibel's Ace is the place.

Jewelry

Cedar Chest Fine Jewelry *(in Tahitian Gardens, 1993 Periwinkle Way, Sanibel; 472-2876)* specializes in vintage jewelry, including fancy yellow diamonds. New pieces feature diamonds and richly colored gemstones, as well as sea-life designs in 14K and 18K gold. Cedar Chest also buys diamonds and old jewelry.

Quality island-themed creations, such as dolphin earrings, are the trademark of **Congress Jewelers** *(in Periwinkle Place, 2075 Periwinkle Way, Sanibel; 472-4177)*. The largest jewelry store on the islands, the family-owned shop has the largest selection. Brands include Rolex, Cartier, Tag Heuer, Raymond Weil, Philippe Charriol, Quadrillion and Mikimoto. They repair watches, too.

Friday's Fine Jewelers *(in Jerry's Shopping Center, 1700 Periwinkle Way at Casa Ybel Rd., Sanibel; 472-1454)* is the home of the Sanibel Diamond, a hand-picked, numbered series. Island-themed jewelry motifs include flip-flop sandals, pail and shovels, and sea turtles. Friday's also offers a wide selection of certified pre-owned Rolex watches and an assortment of diamond dials and bezels for upgrading existing Rolexes. Selections also include Fabergé eggs and Waterford crystal.

Rene's Artisans of Fine Jewelry *(in the Olde Sanibel shops, 630 Tarpon Bay Rd., Sanibel; 472-5544)* offers dazzling diamond and gold jewelry, interesting gemstone pieces and sea-life themes.

Master coin maker Gene Gargiulo creates unique Sanibel treasures at **Sanibel Coin and Jewelry** *(in Bailey's Shopping Center, 2439 Periwinkle Way at Tarpon Bay Rd.; Sanibel; 395-3899)*. Our favorite: his 14K gold sea creatures perched on tiny reefs of freshwater pearls.

Both a jewelry store and a goldsmith's workshop, **Sanibel Island Goldsmith** *(in the Forever Green shops, 2055 Periwinkle Way, Sanibel; 472-8677)* offers handcrafted 14K and 18K gold sea-life jewelry, along with many gold and platinum settings with colored gems.

Bill Wilson creates hand-crafted jewelry at his intimate workshop and showroom, **William E. Wilson Fine Jewelry Design & Diamond Broker** *(in the Village, 2340 Periwinkle Way, Sanibel; 472-8590)*. He specializes in remounts and custom rings, using his 37 years of experience.

Liquor and wine

Manager Joe Suarez makes sure the **Grog Shop** *(in Bailey's Shopping Center, Periwinkle Way at Tarpon Bay Rd.; Sanibel; 472-1682)* carries just about any wine and liquor you could want, including French champagne. The Grog Shop has the islands' only walk-in cigar humidor, stocked with the 15 top-selling brands (Arturo Fuente is the most popular; others include Macanudo, Partagas and Montecristo).

Huxters Liquors *(near the causeway at 1201 Periwinkle Way, Sanibel; 472-3333)* has over 2,000 bottles of wine, champagne, liquor and beer (including micro-brew and imported brands), and imported cigars. Limited delivery is available on Sanibel.

Sanibel Spirits *(in Jerry's Shopping Center, 1700 Periwinkle Way at Casa Ybel Rd., Sanibel; 472-8668)* has the islands' largest selection of single malt scotch. Choose a Merlot, Chardonnay or Cabernet Sauvignon from the fine wine department, or pick up premixed cocktails. The new cigar department carries Macanudo, Partagas and H. Upmann, among others.

Sanibel Wine & Coffee Co. *(in Tahitian Gardens, 2003 Periwinkle Way, Sanibel; 472-1144)* offers a wide selection of, yes, wine and coffee. Co-owner Steve Cearley will help you pick out a nice wine to go with your beach picnic or romantic dinner. Accessories include rice-paper picnic napkins.

Pharmacies

Bailey's Corner Pharmacy *(in Bailey's Shopping Center, Periwinkle Way at Tarpon Bay Rd.; Sanibel; 472-4149)*, a section of Bailey's General Store, fills prescriptions Monday through Friday, 9 a.m. to 5 p.m. See Bailey's listing under Groceries, above.

Barrier Island Pharmacy *(across the street from Jerry's Shopping Center at 1721 Periwinkle Way, Sanibel; 472-8866)* is just that: an old-school, friendly pharmacy straight out of the 1950s, with no magazines, books, candy aisle or other merchandise that clutter today's modern "drug stores." It prides itself on helping visitors who have left their prescription at home.

Newly-remodeled **Eckerd Drugs** *(across from Palm Ridge Place at 2331 Palm Ridge Rd., Sanibel; 472-1719)* has the best prices on 12-packs of soft drinks and beer. The islands' only chain drug store, it offers one-hour photo processing and a great selection of film, batteries and sunscreen.

Seashells

Island Gifts and Shells *(1609 Periwinkle Way, Sanibel; 472-4318)* has a good supply of shells and shell bottles, as well as shell-themed gifts and books.

Neptune's Treasures *(across from Dairy Queen, 1101 Periwinkle Way, Sanibel; 472-3132)* is a find, hidden among some small business condos just west of the causeway. Neptune's has an excellent assortment of shells (including the world-record horse conch), fossils, dinosaur teeth and eggs, Spanish treasure coins, native American artifacts and the largest selection of shark teeth in the region. The passion and care devoted to this small store is evident in the painstakingly hand-printed display notes.

The unassuming **Sanibel Seashell Industries** *(just off Periwinkle Way at 905 Fitzhugh St., Sanibel; 472-1603)* is a seashell conglomerate. Inside it's packed with bins and buckets of shells from Sanibel and around the world, plus specimen shells and preserved sea-life. National decorating magazines, including Martha Stewart Living, buy their supplies here. Owners Bill Strange and Gary Greenplate also sell shell lamps, paraphernalia and sailor's valentine kits.

She Sells Sea Shells *(1157 Periwinkle Way, Sanibel; 472-6991; and 2422 Periwinkle Way, Sanibel; 472-8080)* offers local and worldwide shells, corals and exotic sea-life, shell mirrors, gifts, jewelry, lamps, craft supplies and T-shirts. Ask to see their homemade Christmas ornaments and novelties; they'll even customize items for you.

Word of mouth must be the secret of the **Shell Net** *(in Bailey's Shopping Center, Periwinkle Way at Tarpon Bay Rd.; Sanibel; 472-1702)*. It's hidden back in the corner of Bailey's Shopping Center, but has been in business for 32 years. Island-themed gifts and decorative items fill the shelves, as well as shells, shell craft, ornaments, wood carvings, Beanie Babies and jewelry. All the crafts are made locally — nothing here is stamped "Made in China."

Showcase Shells (in the Heart of the Island shops, 1614 Periwinkle Way, 472-1971) offers specimen shells, aquarium and decorative coral, jewelry, gifts and shell lamps.

Secondhand stores

You'll find rare items at terrific prices at **Designer Consigner** (near Tarpon Bay Rd. at 2460 Palm Ridge Rd., Sanibel; 472-1266). Manager Phyllis Marten focuses on furniture (especially unusual decorator pieces) and offers designer clothing (often new, but selling for half price or less), artwork, lamps, unusual gift items and accessories.

A few doors down is the **Goodwill Boutique** (near Tarpon Bay Rd. at 2440 Palm Ridge Rd., Sanibel; 395-1225). Goodwill Industries calls this store a "boutique" because it stocks higher-quality items than a regular Goodwill store. Clothing for men and women is grouped in easy-to-find sections, including a rack for plus sizes. The bed, bath and kitchenware section is especially popular.

Noah's Ark (behind the Episcopal Church, 2304 Periwinkle Way, Sanibel; 472-3356) is a treasure. The large store is packed with ever-changing donated merchandise from islanders, with all proceeds going to charity. Electronics, clothing, bedding, suitcases — it's all here. The Designer Boutique has the finest of labels, with even the most expensive dresses only about $30. The Barnes and Noah book section has paperbacks and hardbacks. Another area features furniture, large items and artwork. During the season Noah's Ark gets new items every day. But get here early: in-the-know locals arrive when the doors open, and the best finds go right away. In-season hours are 9:30 to noon weekdays.

Toys

Quality toys from Lego, Playmobil and Brio fill most of the shelves at the **Cheshire Cat** (in Tahitian Gardens, 1999 Periwinkle Way,

Sanibel; 472-3545). Other brands include Ty Beanie Babies, Madame Alexander Dolls, Madeline, Thomas the Tank Engine and Hello Kitty. Browse through early learning toys, children's books and music, science and craft kits, kites, outdoor toys and more.

The favorite store of island kids, **Needful Things** (in Tahitian Gardens, 1995 Periwinkle Way, Sanibel; 472-5400) is a groovy collection of kid's stuff — stickers, sports cards, ornaments, retro candy and other irresistible goodies. Brands include Ty Beanie Babies, Betty Boop, Pokemon cards, Dragonball Z, Hello Kitty and Barbie. "I recommend Needful Things because they have useless stuff," says Sanibel School student Kory Phillips. Formerly a retro "junk" shop, it still carries antique jewelry and handkerchiefs.

Toys Ahoy (in Periwinkle Place, 2075 Periwinkle Way, Sanibel; 472-4800) offers many educational games, books, kites, stuffed animals and brands including Playmobil, Madeline, Thomas the Tank Engine and Madame Alexander. A second location upstairs, Toys Ahoy Collectibles, has collector's cards, dolls, bears, tin toys, nesting dolls, Dover books and other retro items.

Video rentals

The **Island Store** (11500 Andy Rosse Ln., Captiva; 472-2374) is the rental source on Captiva. Not a huge selection, but very convenient for Captiva visitors.

Little Nancy's Sunrise Video (near the lighthouse, 359 Periwinkle Way, Sanibel; 472-6364) rents videos, VCRs and camcorders.

Video Scene (in Bailey's Shopping Center, Periwinkle Way at Tarpon Bay Rd.; Sanibel; 472-1158) is the islands' largest video rental store. It also rents VCRs and video games.

Miscellaneous

The **Audubon Nature Store** (in Tahitian Gardens, 1985 Periwinkle Way, Sanibel; 395-2050) sells Bushnell and Bausch & Lomb binoculars, bird feeders and a good variety of nature guides and birding books, as

At the Cheese Nook

well as frames, candles, wind chimes and music. A large children's section has nature-themed toys, books, games and stuffed animals.

Addicted to cheese? **Cheese Nook** (in Periwinkle Place, 2075 Periwinkle Way, Sanibel; 472-2666) has a tempting selection and daily specials. Other standouts include gourmet foods, cookbooks and one of the best selection of hot sauces we've ever seen. We shop here often.

Temptations at **Chocolate Expressions** (in Periwinkle Place, 2075 Periwinkle Way, Sanibel; 472-3837) include rich homemade fudge, Godiva chocolates and hand-dipped chocolate strawberries. Those with willpower may settle for some imported candy, Jelly Belly jellybeans or a fruit smoothie.

Wonderful aromas surround you at **Escentials** (in the Village, 2340 Periwinkle Way, Sanibel; 472-7770). Items include lotions, cosmetics, oils, bath crystals, incense and candles, with dozens of samples to try. Sleepwear and handmade jewelry is also available. As one islander says, "Escentials is 'the bomb'!"

Need a little Christmas in July, March or October? **Forest of Flowers Christmas Shoppe** (in the Heart of the Island shops, 1622 Periwinkle Way, Sanibel; 888-264-0711) offers year-round ornaments, tree hangings and other holiday paraphernalia.

Searching for a wall hanging that says "Cats Rule, Dogs Drool"? You'll find it at **Jenafish** (in the Village, 2340 Periwinkle Way, Sanibel; 472-3098), as well as hundreds of other

unique items for pet lovers, from apparel to food bowls. All have a fun, kicky theme.

Mel Fisher's Sanibel Treasure Company/ Treasures of the Atocha Exhibit and Gift Shop (in the Winds Plaza, 2353 Periwinkle Way, Sanibel; 395-5376) offers original and replica pieces from the shipwrecked Spanish galleon Atocha, including jewelry and silver and gold coins. A museum-quality exhibit shows off $40 million in treasure, including a cannon, coins, plates, bones and a gold bar you can lift yourself to feel its heft. An 8-minute video explains the history of the ship and its cargo. A conservation lab shows how the pieces were cleaned and restored after being lost at sea for hundreds of years.

Sanibel 5 & 10 (at Palm Ridge Place, 2330 Palm Ridge Rd., Sanibel; 472-8288) is an island version of the national chain Spencer's Gifts: Kitsch, giggles, and silly souvenirs, including Hula girls for your dashboard and battery-powered chimps that play cymbals. A "risque business" spot is tucked to the side, and there's a special shelf "for people who appreciate tacky."

Tarpon Bay Recreation Center (900 Tarpon Bay Rd., Sanibel; 472-8900) has the best T-shirts on the islands — the most funky and original. This small, eclectic shop also carries great hats, fanciful wind socks, candles that look exactly like shells, world music, an extensive book section, toys, fishing licenses and tackle, even ice cream.

Terry's Tennis Shop (in the Dunes Golf and Country Club, 949 Sandcastle Rd., Sanibel; 472-3522) is a fully-stocked pro shop, with racquets, tennis clothes, shoes and accessories. Brands include Boast, Descente, Snauwaert and Marcia Originals. They restring rackets, too.

Trudie Prevatt and the other creative women at **Three Crafty Ladies** (in the Heart of the Island shops, 1620 Periwinkle Way, Sanibel; 472-2893) have a shop full of arts and crafts supplies, fabric, notions and shellcraft. They give free craft demonstrations every Tuesday and Thursday during the winter season.

Accommodations

Sanibel and Captiva have many places to stay suitable for all tastes, from primitive cabins to luxurious resorts. Many small hotels line the southern shore of Sanibel. Most were built in the 1960s and early 1970s and retain the look and feel of that era. Don't overlook the properties off the Gulf; many are charming, family-owned operations with relatively bargain rates. Other options include everything from rustic campsites to fully equipped apartments. But don't expect many familiar names; the only franchises on the islands are Holiday Inn and Best Western. Rates on the islands aren't cheap, though prices fluctuate greatly according to the season and location.

Price categories reflect quoted rates per night in high season, including tax and service charges (prices during the summer are often 30 to 50 percent less than in these peak months of February, March and April):

$ under $100
$$ $100-$200
$$$ $200-$300
$$$$ $300-$400
$$$$$ over $400

Superior Small Lodgings. The symbol _SSL_ denotes a property that has been accredited by the Superior Small Lodging program. Each has passed an annual inspection and meets SSL criteria for cleanliness, comfort, privacy and safety. SSL properties have less than 51 rooms.

Sanibel

Anchorage Inn of Sanibel (1245 Periwinkle Way; 395-9688; fax 395-2411. Pool. **$-$$**). This newly renovated inn has hotel rooms, one- and two-room efficiencies and two-bedroom A-frame cottages. All have refrigerators. A-frames have spiral staircases. Within walking distance of stores and restaurants.

Facing page: An Island Inn cottage

Beach Road Inn (764 Beach Rd.; 395-1314 or toll-free 877-501-7600; fax 395-1921. BBQ grills, hot tub, laundry. _SSL_ **$$$**). This four-suite inn is a short walk to the Gulf. Each two-bedroom unit sleeps four. Each has a living room, dining room, kitchen, one and one-half baths and two screened lanais. Videocassette players, washers and dryers and ironing boards provided. A manatee and calf often swim in the canal in the back.

Beachview Cottages (3325 West Gulf Dr.; 472-1202 or toll-free 800-860-0532; fax 472-4720. Ten percent discount for seven-night stay in summer. BBQ grills, laundry, pool. On Gulf. _SSL_ **$$-$$$**). Studio apartments and one- and two-bedroom cottages, each with screened porch. Daily towel service; weekly maid service. Shell and fish service area. Beach sundeck and tiki hut. Cribs available.

Best Western Sanibel Island (3287 West Gulf Dr.; 472-1700; for reservations call toll-free 800-645-6559; fax 481-4947; Germany, Austria and Switzerland 069-44-60-02, fax 069-43-96-31. Packages available. Rates include continental breakfast. BBQ grills, bicycles, daily maid service, laundry, picnic tables, pool, shuffleboard and tennis courts. On Gulf. **$$-$$$$$**). These 45 rooms each have a microwave, refrigerator or kitchenette, and free HBO. Most have screened terraces and garden or Gulf views. Three two-bedroom suites include fully-equipped kitchens. Each room was recently refurbished. Guests get free use of chaise lounges (on the beach) and tennis equipment. Formerly the Jolly Roger.

Blue Dolphin (4227 West Gulf Dr.; 472-1600 or toll-free 800-648-4660; fax 472-8615. BBQ grills, bicycles, laundry. On Gulf. _SSL_ **$$**). These barefoot-casual Gulf-front cottages are run by 20-year island residents. The nine efficiency and one-bedroom units have kitchens with microwaves, individually controlled air conditioning, telephones and sun decks. Guests have complimentary use of lounge chairs, beach umbrellas and bicycles. The Blue Dolphin is the last business along West Gulf Dr., in a quiet area.

Brennen's Tarpon Tale Inn (367 Periwinkle Way; 472-0939 or toll-free 888-345-0939; fax

Casa Ybel Resort

472-6202. *Rates include continental break-fast. Bicycles, laundry. SSL* **$$-$$$**). This 40-year-old, five-bungalow inn was completely renovated in 1996. Each unit has its own theme; all have tile floors, whitewashed wicker and rattan, and antique oak furniture. Landscaped grounds hide the inn from the street. Located near the lighthouse, it's a four-minute walk (or two-minute bike ride) to the beach. Guests get free use of a video library, beach chairs and umbrellas. Ask owners Joe and Dawn Ramsey for advice on restaurants and activities.

Buttonwood Cottages (*1234 North Buttonwood Ln.; 395-9061 or toll-free 877-395-COTTAGE; fax 395-2620. BBQ grills, bicycles, laundry, pool. SSL* **$$-$$$**). These one- and two-bedroom suites, efficiencies and studio cottages have king beds, screened porches, kitchens and videocassette players. Guests have use of beach chairs, towels, umbrellas and fishing equipment. Grounds include hammocks and hot tubs.

Caribe Beach Resort (*2669 West Gulf Dr.; 472-4526 or toll-free 800-330-1593. BBQ grills, bicycles, horseshoe pit, hot tub, laundry, pool, shuffleboard and volleyball courts. Pets 25 lbs. and under allowed. On Gulf. SSL*

$$-$$$). In a shady location on the beach, Caribe offers 19 efficiency apartments, five one-bedroom apartments, a one-bedroom cottage and a two-bedroom cottage. Efficiencies include a kitchen, Murphy bed and a double pullout couch. All apartments have a deck or balcony, private phone, cable TV and videocassette player.

Casa Ybel Resort (*2255 West Gulf Dr.; 472-3145 or toll-free 800-276-4753; fax 472-2109. Packages available. BBQ grills; bicycles; daily maid service; laundry; pool; restaurant; shuffleboard, tennis and volleyball courts. On Gulf.* **$$$$-$$$$$**). Now a modern, 114-unit resort, Casa Ybel has been a major Sanibel destination for over 100 years. Located on 28 Gulf-front acres, it features one- and two-bedroom suites (all beachfront), each with a sleeper sofa in the living room and a fully equipped kitchen. For families, the resort has a children's play area, a children's pool and a Kid's Club, with programs on shelling, the beach, nature and treasure hunts. Classes teach guests to make beach bags and elaborate sand castles. The Olympic-sized main pool has an adjacent whirlpool.

Castaways at Blind Pass (*6460 Sanibel-Captiva Rd.; 472-1252 or toll-free 800-375-0152; fax 472-1020. Ten percent discount for seven-night stay during summer months. Packages available. BBQ grills, bicycles, laundry, pool. Three-night minimum stay during holiday periods. Some pets OK. Some units on Gulf.* **$$-$$$$**). At the west end of Sanibel, the Castaways sits on a small bay off Blind Pass, with some cottages on the Gulf. It includes water-view cottages, efficiencies and motel units, all with cable TV, some with kitchens. Daily towel service; weekly maid service. Public telephones available. The Castaways has a shell and fish cleaning area; the adjacent marina offers boat dockage, rentals, charters, bait and sundry items.

Colony Resort (*419 East Gulf Dr.; 472-5151 or toll-free 800-342-1704. BBQ grills, laundry, pool. Three-night minimum stay during holidays. On Gulf.* **$$**). The Colony features single and duplex cottages and one-bedroom condos, which sleep four. Linen service is complimentary; maid service optional. All units have a kitchen; condos have screened porches. The complex also has picnic tables.

Driftwood Inn (711 Donax St.; 395-8874; fax 472-6935. Some pets allowed. **$$**). These efficiency, one-bedroom, and two-bedroom cottages have a kitchen with cookware and utensils, living room, dining area and screened-in porch with a dining table. One-bedrooms accommodate four people; two-bedrooms sleep six. Bedrooms have queen-size beds; living rooms queen-size sofa beds. Beach chairs, umbrellas and linens (except for beach towels) are provided.

Forty-Fifteen Resort (4015 West Gulf Dr.; 472-1232. Weekly rentals only. Laundry, pool, tennis courts. Not handicap accessible. Non-smoking. On Gulf. SSL **$-$$$**). Forty-Fifteen has nine vacation cottages on the uncrowded end of West Gulf. Seven accommodate four people, two sleep six, one sleeps two. Kitchens have microwaves, dishes, pots and pans. Complimentary linens.

Gulf Breeze Cottages (1081 Shell Basket Ln.; 472-1626 or toll-free 800-388-2842; fax 472-4664. Laundry, shuffleboard courts. On Gulf. SSL **$$-$$$$**). Recently remodeled, these 13 one-room efficiencies and one- and two-bedroom cottages are at the end of a private lane. Each has a kitchen and refrigerator. Cottage No. 7 has a two-way view of the Gulf. Babysitting available.

Holiday Inn Beach Resort (1231 Middle Gulf Dr.; 472-4123. Bicycles, pool, restaurant, tennis courts. On Gulf. **$$$**). All 98 rooms include two double beds or one king-size bed, small refrigerator, iron and ironing board, coffee maker, hair dryer and in-room safe. Pool-side tiki snack bar, gift shop. Complimentary beach cabanas.

Hurricane House Resort (2939 West Gulf Dr.; 472-1696 or toll-free 800-448-2736; fax 472-1718. BBQ grills, bicycles for rent, laundry, pool, tennis courts. On Gulf. **$$$-$$$$$**). These two-bedroom townhouses feature kitchens, videocassette players and screened balconies. The grounds have two whirlpool spas and picnic areas. Guests receive complimentary tennis privileges and greens fees at the Dunes Golf & Tennis Club.

Island Inn (3111 West Gulf Dr.; 472-1561 or toll-free 800-851-5088; fax 472-0051. Two-night minimum for efficiency units and for all units on weekends. Daily maid service; laundry; pool; restaurant; table tennis; croquet, shuffleboard, tennis, volleyball courts. Butterfly garden. On Gulf. **$$-$$$$$**). Established in 1895, this genteel inn consists of three lodges and several cottages in a tranquil setting. All units have refrigerators. The inn operates on a Modified American plan (breakfast and dinner included in the room rate) from Nov. 15 to May 1. The "blowing of the conch" summons guests to dinner (formal, jackets for men) during the season.

Kona Kai Motel (1539 Periwinkle Way; 472-1001 or toll-free 800-820-2385. Rates include continental breakfast. BBQ grills, bicycles, pool. Some pets allowed. **$**). The Kona Kai has motel rooms and efficiencies, and suites that sleep six. Suites include a living room, kitchen, bedroom. Complimentary linen service. Within walking distance of stores and restaurants. Sanibel River in back. Canoes available. The beach is 3/4 mile away.

Lighthouse Resort & Club (210 Periwinkle Way; 472-4526 or toll-free 800-456-0009; fax 472-0079. Weekly rentals only. BBQ grills, bicycles, laundry, pool, shuffleboard and tennis courts. SSL **$$-$$$**). On the bay, the Lighthouse Resort has 1,800-square-foot apartments. Each has a king-size bed, two twin beds, a pull-out couch, a kitchen and a washer and dryer.

Mitchell's Sand Castles By The Sea (3951 West Gulf Dr.; 472-1282. No credit cards. Laundry, pool. Some pets allowed. On Gulf. **$$-$$$**). Owned by islander Roxanne Palmer, these relaxed one- to four-bedroom cottages have kitchens and screened porches.

Ocean's Reach Condominium (2230 Camino del Mar Dr.; 472-4554 or toll-free 800-336-6722. Weekly rentals only. BBQ grills; bicycles; covered parking; laundry; picnic tables, pool; basketball, shuffleboard and tennis courts. On Gulf. **$$-$$$**). Ocean's Reach offers one-bedroom, one-bath and two-bedroom, two-bath condo units, all with a lanai facing the Gulf. Each has a washer, dryer, dishwasher and videocassette player.

The Palms of Sanibel

Palm View Motel *(706 Donax St., 472-1606; fax 472-6733. BBQ grills, laundry. Some pets allowed. $-$$)*. Motel rooms, efficiencies and one- and two-bedroom apartments all have kitchens (except two motel rooms, which connect and can be rented together). Picnic area; shell and fish cleaning facilities. One block from the beach.

The Palms of Sanibel *(1220 Morningside Dr.; 395-1775 or toll-free 877-749-5093; fax 395-3379. Weekly rentals only. BBQ grills, bicycles, laundry, pool. SSL $$-$$$)*. Located near the lighthouse, these cottages were recently remodeled. Each has ceramic-tile floors, a kitchen with breakfast bar, a queen-size bed, queen-sleeper sofa, two TVs, a videocassette player, screened-in porch and deck. Fresh towels daily. Complimentary umbrellas, beach towels, chairs.

The Parrot Nest Old-Sanibel Resort *(1237 Anhinga Ln.; 472-4212. BBQ grills, daily maid service, laundry. SSL $$)*. Talking parrots are the attraction of the Parrot Nest, a six-room inn near the lighthouse. Rick Flanagan has equipped each room with a refrigerator, microwave, stove and patio. Each sleeps up to three with one king-size bed and one single bed. A cafe, deli and shops are next door. The Gulf is a five-minute walk away.

Pelicans Roost Condominium *(605 Donax St.; 472-2996 or toll-free 877-757-6678; fax 472-0317. Weekly rentals only. No credit cards. BBQ grills, horseshoe pit, laundry, pool, shuffleboard and tennis courts. On Gulf. $$$-$$$$)*. This 3½-acre property has 21 two-bedroom, two-bath apartments each with a screened porch overlooking the Gulf.

Periwinkle Cottages *(1431 Jamaica Dr.; 472-1880; fax 472-5567. No walk-ins accepted; some pets allowed. SSL $$)*. Not on Periwinkle Way, the Periwinkle Cottages are out toward the west end of Sanibel, off Sanibel-Captiva Rd. Cottages have recently been remodeled with new furniture and beds. Each has a screened porch, videocassette player, hair dryer, dinnerware and appliances. Complimentary linen service. Beach chairs, umbrellas, floats, beach towels furnished. Grounds have a secluded pond and gazebo, and climbing equipment for small children. The Sanibel Recreational Complex, with a pool, tennis courts and other amenities, is a short bike ride away.

Periwinkle Park and Campground *(1119 Periwinkle Way; 472-1433. Children under 6 free. No credit cards. BBQ grills, laundry, picnic tables. Some pets allowed, but not dogs. $)*. Periwinkle Park accepts trailers, motor homes, truck campers, tent campers, tents and vans. It has electricity, sewer and water hookups, as well as restrooms and showers. Ice and LP gas are available. The park also has a collection of exotic animals, including tropical birds, monkeys and miniature deer. Attracting an upscale crowd, Periwinkle Park has guests who've been coming annually for decades. About a dozen people live here year-round. The Sanibel River runs along the back of the property. Periwinkle Park was built in 1964 by Albert Muench, who still runs it with his sons, Dick (who co-owns the Periwinkle Lazy Flamingo) and Jerry (a former Sanibel mayor).

Pointe Santo de Sanibel *(2445 West Gulf Dr.; 472-9100 or toll-free 800-824-5442; fax 472-0487. Weekly rentals only. BBQ grills, hot tub, laundry, pool, shuffleboard and tennis courts. On Gulf. $$$-$$$$$)*. These one- to three-bedroom condos each have a breakfast area, dining room, kitchen and a screened lanai viewing the Gulf. Some have rooftop sun decks. All have washers and dryers. The complex has tiki huts, on-site guest service and an activity program.

Sandalfoot Condominium *(671 East Gulf Dr.; 472-2275 or toll-free 800-725-2250; fax 472-5135. Minimum stay three nights. Laun-*

dry; pool; basketball, shuffleboard and tennis courts. On Gulf. **$$-$$$**). All one- and two-bedroom Gulf-front or Gulf-view condos here have a kitchen with microwave, a screened balcony or patio and their own decor. Fresh towels are available weekly.

Sanddollar Condominium (1785 Middle Gulf Dr.; 472-5021 or toll-free 800-794-3107; fax 466-0514. Minimum stay: one week during the winter season; two weeks rest of the year. BBQ grills, laundry, pool, tennis courts. On Gulf. **$$$$-$$$$$**). All 36 two- and three-bedroom units here have a washer and dryer and a screened lanai.

Sandpiper Inn (720 Donax St.; 472-1529 or toll-free 877-227-4737; fax 472-0967. BBQ grills, bicycles, laundry, picnic tables. SSL **$$**). These newly remodeled one-bedroom, one-bath suites each have a living room, dining room and kitchen. Complimentary beach towels and beach chairs.

Sandy Bend (3057 West Gulf Dr.; 472-1190; fax 472-3057. Weekly rentals only. Daily maid service, laundry, tennis courts. On Gulf. **$$$**). Eight individually decorated, two-bedroom apartments have a living and dining room, kitchen with dishwasher, and screened Gulf-front porch. Complimentary linens.

Sanibel Arms Condominium (805 East Gulf Dr.; 472-2259 or toll-free 800-806-7368; fax 472-2420. Weekly rentals only. Extended-stay discounts available. No credit cards. BBQ grills, laundry, pool, shuffleboard courts. On Gulf. **$$$**). Condos have a kitchen, individually controlled air-conditioning, screened porch or balcony. Complimentary linen service. The complex includes a clubhouse with library, beach shower and dockside fish-cleaning facilities. Canal boat dock.

Sanibel Arms West Condominium (827 East Gulf Dr.; 472-1138 or toll-free 800-950-1138. Minimum stay four nights. Laundry, pool, tennis courts. On Gulf. **$$-$$$**). These 13 two-story units each have over 1,000 square feet, with two bedrooms, two baths, a living room, dining room, electric kitchen, balcony and screened-in porch.

Sanibel Beach Club I and II (626 Nerita St. and 265 Periwinkle Way; toll-free 800-456-

0009. Weekly rentals only. BBQ grills; bicycles; horseshoe pit; hot tub and sauna; laundry; pool; basketball, shuffleboard, tennis and volleyball courts. On Gulf. **$$-$$$**). Two-bedroom, two-bath condos each include a kitchen with microwave, linens and dishes, washer and dryer, videocassette player and screened-in porch. The 6.3-acre complex also includes a children's play area.

Sanibel Inn (937 East Gulf Dr.; 472-3181 or toll-free 800-965-7772; fax 481-4947. BBQ grills, bicycles, daily maid service, pool, restaurant, room service, tennis courts. On Gulf. **$$-$$$$$**). Most of these 96 units are hotel rooms that offer a king or two queen beds, refrigerator, microwave, a screened balcony or patio, coffee service and a videocassette player. One-bedroom suites have king-size beds and queen sleepers. Two-bedroom suites have a kitchen with dishwasher, refrigerator, stove and microwave, and living room with sleeper (these units sleep six, with a master bedroom with two double beds or a king-size bed, and a second bedroom with two twins). Free daily newspaper, HBO and Disney Channel. Poolside cabana, on-site rentals of umbrellas, cabanas, floats and kayaks; free rental of tennis gear. Guests can attend Florida native-gardening classes. Other activities include Beach Discovery, Sanibel Shell Safaris, botanical garden socials and dolphin watches. Children's programs include "How Does Your Garden Grow," "Bugs Don't Bug Me" and Mad Hatter Tea Parties. Craft projects include decorating visors, T-shirts and picture frames. The inn received the 1999 Business Partner Award from the Sanibel Captiva Conservation Foundation. The 8-acre grounds are landscaped to attract butterflies and hummingbirds, and have over 500 palms.

Sanibel Moorings Condominium (845 East Gulf Dr.; 472-4119 or toll-free 888-FLA-ISLE; Germany and U.K 800-237-5144. One-week minimum stay during the winter season. Nightly rentals off season. BBQ grills, laundry, pool, tennis courts. On Gulf. **$$-$$$$**). Located between the Gulf and a lagoon, this six-acre resort has a lagoon boat dock. Complimentary beach chairs and lounges. A business center has a PC and fax.

Sanibel Siesta *(1246 Fulgur Street; 472-4117 or toll-free 800-548-2743; fax 472-6826. Minimum stay three nights. Laundry, pool, shuffleboard and tennis courts. On Gulf. $$$).* All 64 two-bedroom, two-bath units here have a living room, dining area, kitchen with microwave and screened lanai. The Beachview Country Club is within walking distance.

Sea Shells of Sanibel Condominium *(2840 West Gulf Dr.; 472-4634 or toll-free 800-533-4486; fax 472-0724. Minimum stay one week during winter, three days rest of year. BBQ grills, laundry, pool, shuffleboard and tennis courts. $$).* These two-bedroom, two-bath units are divided into Group A (1,050 square feet) and Group B units (1,350 square feet). Each has a kitchen with dishwasher, garbage disposal, self-cleaning oven and microwave. All have telephones and screened porches; some have videocassette players and stereo receivers. On-site management.

Seahorse Cottages *(1223 Buttonwood Ln. North; 472-4262; fax 466-6149. No children. BBQ grills, bicycles, laundry, pool. SSL $$-$$$).* Each cottage has a living room, kitchen, bedroom, porch, private front and back entrance, antique oak furnishings, paddle fans, ceramic-tile floors, individually controlled air conditioning and videocassette player. Kitchens have solid-carbon water filters. Grounds include a shell-cleaning table and hammock. Complimentary beach towels, chairs and umbrellas. A few blocks from the Gulf; 200 feet from the bay.

Seaside Inn *(541 East Gulf Dr.; 472-1400 or toll-free 800-831-7384; fax 481-4947. Rates include continental breakfast. BBQ grills, bicycles, laundry, picnic tables, pool, restaurant, shuffleboard courts. On Gulf. $$$-$$$$).* This 33-unit inn has one-, two-, and three-bedroom cottages with kitchens, separate dining areas and living rooms and videocassette players. Poolside studios offer small refrigerators and microwaves. Some units have private porches or balconies. Complimentary book and video library.

Shalimar Resort *(2823 West Gulf Dr.; 472-1353 or toll-free 800-995-1242. Ten percent discount to AAA and AARP members. Packages available. Laundry, pool, basketball and*

shuffleboard courts. *SSL $$-$$$$).* Thirty-three one- and two-bedroom cottages and motel efficiencies feature kitchens with microwaves and furnished linens.

Signal Inn *(1811 Olde Middle Gulf Dr.; 472-4690 or toll-free 800-992-4690; fax 472-3988. Minimum stay three nights. Hot tub, laundry, pool, sauna, air-conditioned racquetball court. Some pets OK. SSL $$-$$$$).* All 19 elevated one- to four-bedroom units have three-sided exposures. Each has a screened porch, kitchen, videocassette player, washer and dryer and parking underneath. Gazebo; outdoor foot showers. On a dead-end road 200 yards from the beach.

Song of the Sea *(863 East Gulf Dr.; 472-2220 or toll-free 800-231-1045; fax 472-8569; Germany, Austria and Switzerland 069-44-60-02, fax 069-43-96-31. Rates include outdoor continental breakfast. BBQ grills, bicycles, laundry, pool, shuffleboard and tennis courts, whirlpool. On Gulf. $$$-$$$$$).* Song of the Sea offers 10 pool-view rooms, 12 Gulf-view rooms and eight one-bedroom Gulf-front suites. Each has a kitchen, dining area and screened terrace. Pool-view rooms have two queen-size beds; Gulf-view rooms have one king-size bed; one-bedroom suites have one king-size bed and a queen-size living room sleeper sofa. All are nonsmoking. Guests get a bottle of wine, fresh-cut flowers, bottled water, daily newspapers and use of the inn's book and video library, videocassette players, beach umbrellas and chaise lounges. Popular with Europeans.

Sundial Beach Resort *(1451 Middle Gulf Dr.; 472-4151 or toll-free 800-965-7772. BBQ grills, bicycles, fitness room, laundry, pools, restaurants, tennis courts, whirlpool. On Gulf. $$$-$$$$$).* The 33-acre, 270-unit Sundial has studio, one-bedroom and two-bedroom condos with kitchens. All have a dining area and living room; some have a den and/or a screened balcony or patio. All have in-room safes. The resort rents catamarans, sea kayaks and boogie boards. The 12-court tennis facility holds tournaments among the guests. Private lessons and group clinics available. Organized activities for children, teenagers, adults and groups. Kids camps

Sundial Beach Resort

are available for half days, full days and evenings. Kid's activities can include shell crafts, tennis clinics, aquacize and hermit-crab races. Environmental Discovery Programs include lectures and videos; hands-on learning in activities such as a "beach walk and talk." A business center offers PCs, laser printing, a typing service and cellular phone and pager rentals.

Sunshine Island Inn *(642 East Gulf Dr.; 395-2500. BBQ grills, laundry, pool.* **$$-$$$***).* This family-operated five-room inn is across the street from the beach. Each room has been recently renovated with Mexican tile and Berber carpet, a sliding glass door that leads out to the pool, a videocassette player and either an efficiency or full kitchen.

Surfrider Beach Club *(555 East Gulf Dr.; 472-2161. Two-night minimum stay. BBQ grills, bicycles, hot tub, laundry, pool, shuffleboard and tennis courts. On Gulf.* **$-$$$***).* The Surfrider has 30 small one-bedroom condos and one two-bedroom unit, all with microwaves and videocassette players. Complimentary beach chairs, umbrellas.

Tropical Winds Motel *(4819 Trade Winds Dr.; 472-1765. Pool. On Gulf.* **$$$-$$$$$***).* Six units are on the beach (four more a short

walk away) at this old-time Sanibel property, located in a quiet area. All units are efficiencies, with coffee makers and dishes.

Waterside Inn On The Beach *(3033 West Gulf Dr.; 472-1345 or toll-free 800-741-6166. Wedding and family reunion packages available. BBQ grills, bicycles, laundry, pool, shuffleboard courts. Some pets OK. On Gulf.* **$$-$$$$***).* Nestled among palms, the Waterside Inn has rooms, efficiencies, cottages and condos. Cottages have kitchens and dining nooks. Bedrooms have king-size beds, dressers and large closets. Efficiencies have queen beds, kitchens with microwaves and private patios or balconies. A honeymoon cottage has a double Jacuzzi and a fireplace. Condos are surrounded by trees. Non-smoking rooms available. Beach umbrellas can be rented. Formerly the Snook Motel.

Westend Paradise of Sanibel *(1389 Tahiti Dr.; 472-9088; fax 472-8009. Two-bedroom suites have a two-night minimum stay. During holidays winter rates apply with a three-night minimum. BBQ grills, bicycles, laundry. SSL* **$-$$***).* This small inn is in a quiet subdivision 1,000 feet from a secluded beach. One- and two-bedroom suites include kitchens, individually controlled air conditioning, telephones. The garden has shady gazebos. Owners Carola and Peter Wilkens provide beach chairs and umbrellas.

West Wind Inn *(3345 West Gulf Dr.; 472-1541 or toll-free 800-824-0476; fax 472-8134; U.K. toll-free 00-800-897-44321, U.K. toll-free fax 0800-9625-67; Germany toll-free fax 0130-810990. Group rates available. Minimum stays required during certain holiday and peak-season periods (call for details). Bicycles, laundry, pool, restaurant, tennis courts. On Gulf.* **$$$***).* All rooms have refrigerators, data ports and screened lanais; many have kitchens. Driftwood Building rooms are larger; include upgraded appointments and videocassette players. Non-smoking rooms have king beds or two double beds. Complimentary cribs with advance notice. Beach games and rentals. Pool area includes bar, splash pool for toddlers. Free tennis, tennis clinics. Complimentary 24-hour coffee and tea; daily newspapers. Gift shop, concierge service. On the quiet end of West Gulf Dr.

Captiva

Captiva Island Inn *(11509 Andy Rosse Ln.; 395-0882 or toll-free 800-454-9898; fax 395-0862. Rates include breakfast. Bicycles. SSL $$$).* Down the street from the beach, this B&B is next to shops, restaurants and art galleries. Six one- and two-bedroom cottages each have a living room with a sleeper sofa, kitchen and bath. Complimentary beach chairs and sunscreen.

Jensen's On The Gulf *(15300 Captiva Dr.; 472-4684. Minimum stay may be required. On Gulf. $$$-$$$$$).* Nine shady beach homes and apartments.

Jensen's Twin Palm Resort *(15107 Captiva Dr.; 472-5800. Minimum stay may be required. SSL $$).* This historic bayside retreat has 14 cottages and apartments. The Twin Palms marina is adjacent.

Maddison Suites *(11508 Andy Rosse Ln.; 472-3113 or toll-free 800-472-0638. Weekly rates available. Breakfast included. Bicycles. Some pets OK. SSL $$-$$$).* Located in the Old Captiva Village, this four-suite B&B is a two-minute walk from the beach. Suites sleep two to four. Shops, restaurants nearby.

South Seas Resort *(Captiva Dr.; 472-5111 or toll-free 800-965-7772; fax 481-4947. Packages available. BBQ grills, bicycles, daily maid service, 9-hole golf, laundry, pools, restaurants, tennis courts, water-sports center. On*

Jensen's Twin Palm Resort

Gulf. $$-$$$$$). This 330-acre resort covers the north end of Captiva. Over 600 accommodations include hotel rooms; one-, two- and three-bedroom condos; cottages; and private homes with their own pools and tennis courts. Complete marina. Twenty-one tennis courts; 18 swimming pools. Offshore Sailing School. Free trolley service.

'Tween Waters Inn *(15951 Captiva Dr.; 472-5161 or toll-free 800-223-5865; fax 472-0249. Packages available. Rates include continental breakfast buffet. Rates based on occupancy of 1–2 in rooms and efficiencies and 1–4 in suites. Daily maid service, fitness center, laundry, pool, restaurants, tennis courts. $$-$$$$$).* On a narrow spit of land between the Gulf and Pine Island Sound, this inn has a bayside marina in back and sandy beaches in front across the street. Three restaurants, poolside bar and grill, free tennis and tennis clinics, and (extra-cost) private lessons. The marina has boat, canoe and kayak rentals. Fishing, shelling, sailing and kayaking guides available. Clothing store. Anne Lindbergh wrote her classic book, "Gift from the Sea," while staying here.

Rental agencies

Rental agencies represent the owners of private homes and condos available for rent. Each manages a large number of properties. You tell the agency the dates you will be visiting; it matches you up with a rental.

- **1-800-Sanibel**
  2000 Periwinkle Way, Sanibel;
  472-1800, 1-800-SANIBEL
- **Captiva Island Realty**
  11514 Andy Rosse Ln., Captiva; 472-3158
- **Cottages to Castles**
  2427 Periwinkle Way, Sanibel; 472-6385
- **Gopher Enterprises**
  P.O. Box 186, Sanibel; 472-5021
- **Grande Island Vacations**
  1506 Periwinkle Way, Sanibel; 472-5322
- **Island Vacations**
  1101 Periwinkle Way, Sanibel; 472-7277
- **Kenoyer Real Estate Corp.**
  2669 West Gulf Dr., Sanibel; 472-4526
- **Makin' Waves in Paradise**
  2402 Palm Ridge Rd. #4, Sanibel; 395-1811

MERISTAR

■ **North Captiva Island Club Resort**
P.O. Box 1000, Pineland; 395-1001
■ **Priscilla Murphy Vacation Rentals**
1177 Causeway Rd., Sanibel; 472-4883
■ **Properties in Paradise**
1101 Periwinkle Way, Sanibel; 472-4104
■ **ReMax of the Islands Rentals**
2400 Palm Ridge Rd., Sanibel; 472-5050
■ **Reservation Central**
695 Tarpon Bay Rd. #1, Sanibel; 395-3682
■ **Royal Shell Vacation Properties**
1200 Periwinkle Way, Sanibel; 472-9111
■ **Sanibel Accommodations**
2341 Palm Ridge Rd., Sanibel; 472-3191
■ **Sanibel-Captiva Central Reservations**
1633B Periwinkle Way, Sanibel; 472-0457
■ **Sanibel Holiday**
1630A Periwinkle Way, Sanibel; 472-6565
■ **Sanibel One**
1633G Periwinkle Way, Sanibel; 395-2610
■ **Sanibel Shores Resort Rentals**
1214 Junonia St., Sanibel; 472-4675
■ **VIP Realty Rental Division**
1560 Periwinkle Way, Sanibel; 472-1613

Off the islands

These four places to stay are just minutes from the Sanibel Causeway:

Country Inns & Suites (13901 Shell Point Plaza, Ft. Myers; 454-9292. Rates include con-

South Seas Resort

tinental breakfast. Daily maid service, heated pool, workout room. **$$-$$$**). Opened in the fall of 2001, this 112-room facility has special rates for extended stays. Each room has a coffee maker and hair dryer. At the entrance to the Shell Point retirement village.

Hampton Inn & Suites (11281 Summerlin Square Dr., Ft. Myers; 437-8888. Rates include continental breakfast. Daily maid service, heated pool. Nonsmoking rooms available. **$$-$$$**). A new offering from this well-respected chain, opened in 2001.

Radisson Inn/Sanibel Gateway (20091 Summerlin Rd., Ft. Myers; 466-1200. Daily maid service, hot tub/Jacuzzi, pool, restaurant, room service. **$$-$$$**). Rooms come with a king-size bed and a double bed or sofa sleeper. Each has a small refrigerator, coffee maker, iron, ironing board and hair dryer. Renovated in the summer of 2001.

Sanibel Harbour Resort & Spa (17260 Harbour Pointe Dr., Ft. Myers; 466-4000. Daily maid service, pool, restaurant, spa, tennis courts, watersports. Many extras. On San Carlos Bay. **$$$$-$$$$$**). This internationally famous resort has 417 guest rooms, suites and condos. Special program for kids.

SANIBEL
COMMUNITY
CHURCH
SUNDAY WORSHIP
8:00 COMMUNION
9:00 CONTEMPORARY
10:30 TRADITIONAL

1740

Resources

Important phone numbers

Emergencies	911
Sanibel Police Department	472-3111
Lee County Sheriff (Captiva)	477-1200
Sanibel Fire Department	472-5525
Captiva Fire Department	472-9494
HealthPark of the Islands	395-1414
C.R.O.W. (for injured animals)	472-3644
Florida Highway Patrol	278-7100
Florida Marine Patrol	332-6966
U.S. Coast Guard	463-5754
Lee Memorial Hospital	332-1111
Poison Control	800-282-3171
AAA 24-hour service	800-365-0933
Local directory inquiries	411
International directory inquiries	00
International operator assistance	01

For long-distance information: dial 1, then the appropriate area code, then 555-1212. **For direct-dial calls** to another area code: dial 1, the area code and then the 7-digit number. For international calls dial 011, the country code (Australia 61, New Zealand 64, UK 44), then the local area/city code (minus the first 0) and the number.

Florists

- **Floral Artistry**
2400 Palm Ridge Rd., Sanibel; 472-3040
- **Flower Shop of the Islands**
2449 Periwinkle Way, Sanibel; 472-3707
- **Periwinkle Florist & Gift Baskets**
1719 Periwinkle Way, Sanibel; 472-3125
- **Weeds and Things**
2330 Palm Ridge Rd., Sanibel; 472-2112

Gasoline and auto service

- **Amoco ServiceCenter**
1015 Periwinkle Way, Sanibel; 472-2125
- **Hess Express**
2499 Palm Ridge Rd., Sanibel; 472-2198
No auto service
- **Island Garage**
1609 Periwinkle Way, Sanibel; 472-4318
No gasoline

Facing page: Sanibel Community Church

- **Sanibel Shell**
2435 Periwinkle Way, Sanibel; 472-2012
- **7-Eleven**
2460 Periwinkle Rd., Sanibel; 472-8696
No auto service

Hair and nail salons, day spas

- **Beverly Hills Hair Design**
2340 Periwinkle Way, Sanibel; 395-3116
- **Cape Nails**
2407 Periwinkle Way, Sanibel; 472-4145
- **Harry Ruby Salon**
975 Rabbit Rd., Sanibel; 395-0910
- **Island Winds Coiffures**
695 Tarpon Bay Rd., Sanibel; 472-2591
- **New Spirit Hair Design**
630 Tarpon Bay Rd., Sanibel; 472-2371
- **Pat's Hair Kair**
2248 Periwinkle Way, Sanibel; 472-2425
- **Sanibel Barber Shop**
2467 Periwinkle Way, Sanibel; 472-5626
- **Sanibel Beauty Salon**
2467 Periwinkle Way, Sanibel; 472-1111
- **Sanibel Day Spa**
2075 Periwinkle Way, Sanibel; 395-2220
- **Scarlett O'Hair's Beauty Salon**
1711 Periwinkle Way, Sanibel; 472-5699

Libraries

- **Captiva Memorial Library**
11560 Chapin Ln., Captiva; 472-2133
- **Sanibel Public Library**
770 Dunlop Rd., Sanibel; 472-2483

Massage therapists

- **New Spirit Hair Design**
630 Tarpon Bay Rd., Sanibel; 472-2371
- **Sanibel Beauty Salon**
2467 Periwinkle Way, Sanibel; 472-1111
- **Sanibel Day Spa**
2075 Periwinkle Way, Sanibel; 395-2220
- **Sanibel Wellness**
1717 Periwinkle Way, Sanibel; 395-1100

Medical services

- **Auditory Associates Hearing Center**
2418 Palm Ridge Rd., Sanibel; 395-1700
- **Coral Veterinary Clinic**
1530 Periwinkle Way, Sanibel; 472-8387

■ **Drs. Eyecare Centers**
1571 Periwinkle Way, Sanibel; 472-4204
■ **Eye Centers of Florida**
1723 Periwinkle Way, Sanibel; 395-1999
■ **HealthPark of the Islands**
1699 Periwinkle Way, Sanibel; 395-1414
■ **Island Chiropractic Center**
2400 Palm Ridge Rd., Sanibel; 472-9830
■ **San-Cap Medical Center**
4301 Sanibel-Captiva Rd., Sanibel; 472-0700
■ **Sanibel Chiropractic**
1717 Periwinkle Way, Sanibel; 472-0900

Movie theater

■ **Island Cinema**
535 Tarpon Bay Rd., Sanibel; 472-1701

Newspapers and magazines

■ **Captiva Current**
2340 Periwinkle Way, Sanibel; 472-1580
■ **Island Reporter**
2340 Periwinkle Way, Sanibel; 472-1587
■ **Island Sun**
1640 Periwinkle Way, Sanibel; 395-1213
■ **Sanibel-Captiva Islander**
695 Tarpon Bay Rd., Sanibel; 472-5185
■ **Times of the Islands magazine**
1630 Periwinkle Way, Sanibel; 472-0205

Photo processing

■ **Arundel's Hallmark Shoppe**
1626 Periwinkle Way, Sanibel; 472-0434
14900 Captiva Dr., Captiva; 395-0434

Captiva mail box

■ **Eckerd Drugs**
2331 Palm Ridge Rd., Sanibel; 472-0085
■ **MotoPhoto & Portrait Studio**
1700 Periwinkle Way, Sanibel; 472-4414

Postal services

■ **Sanibel Post Office**
650 Tarpon Bay Rd., Sanibel; (800) 275-8777
8:30 a.m.–5 p.m. Mon.–Fri.; 10 a.m.–noon Sat.
■ **Captiva Post Office**
Captiva Dr.; 472-1674
9 a.m. to 4 p.m. Mon.–Fri.
■ **Jerry's Supermarket** (contract office)
1700 Periwinkle Way, Sanibel; 472-9300
8 a.m.–5 p.m. Mon.–Fri.; 8–10 a.m. Sat.

Printers

■ **Big Red Q Quickprint Center**
1101 Periwinkle Way, Sanibel; 472-2121
■ **Island Graphics**
459 Periwinkle Way, Sanibel; 472-4437
■ **Sanibel Print and Graphics**
2400 Palm Ridge Rd., Sanibel; 472-4592

Radio

■ **Adult Contemporary**
WINK-FM, 96.9, Hot adult contemporary
WXKB-FM, 103.9, Adult current hits
■ **Christian**
WAYJ-FM, 88.7, Christian hit radio
WSOR-FM, 90.9, Christian life
■ **Classical/Jazz**
WGCU-FM, 90.1, NPR/classical/jazz/news
WDRR-FM, 98.5, Smooth jazz
■ **Country**
WIKX-FM, 92.9, Country
WWGR-FM, 101.9, Country
WCKT-FM, 107.1, Country
■ **Easy listening/Standards**
WAVV-FM, 101.1, Easy listening
WJST-FM, 106.3, Adult standards
WKII-AM, 1070, Adult standards
■ **Kids**
WMYR-AM, 1410, Radio Disney
■ **News talk**
WINK-AM, 1200, News talk
WTLQ-AM, 1240, Talk
■ **Oldies**
WOLZ-FM, 95.3, '50s-'60s Oldies
WJGO-FM, 102.9, '70s-'80s Oldies

■ Pop
WCCL-FM, 97.7, '80s hits
WKFF-FM, 100.1, Hit music
WQNU-FM, 105.5, Urban hot radio
■ Rock
WARO-FM, 94.5, Classic rock
WRXK-FM, 96.1, Classic rock
WJBX-FM, 99.3, Alternative

Religious services
Services are year-round except as indicated

■ **Bat Yam Temple of the Islands.** *Services at the Sanibel Congregational United Church of Christ, 2050 Periwinkle Way, Sanibel; 472-6684.* Reform services Fri. at 8 p.m.
■ **Captiva Chapel by the Sea.** *11580 Chapin Ln., Captiva; 472-1646.* Services Sun. at 11 a.m Nov.–April.
■ **St. Isabel Catholic Church.** *3559 Sanibel-Captiva Rd., Captiva; 472-2763.* Mass Mon.–Fri. at 8:30 a.m., Sat. at 5:30 p.m.; Sun. at 8:30 a.m. and 10:30 a.m. Confession Sat. at 9 a.m.
■ **St. Michael's and All Angels Episcopal Church.** *2304 Periwinkle Way, Sanibel; 472-2173.* Services Advent–Easter Sat. at 5 p.m. and Sun. at 7:30 a.m., 9:30 a.m. and 11:30 a.m.; Easter–Advent Sun. at 7:30 a.m. and 9:30 a.m. Other services Wed. at 9 a.m.; Thur. at 7:30 a.m. Church school Sun. mornings; youth group Wed. evenings; adult Bible Study Thur. mornings.
■ **Sanibel-Captiva First Church of Christ Scientist.** *2950 West Gulf Dr., Sanibel; 472-8684.* Services Sun. at 10:30 a.m.; Wed. at 7:30 p.m. Sunday school Sun. at 10:30 a.m. Reading room Mon., Wed., Fri. 10 a.m–noon.
■ **Sanibel Community Church.** *1740 Periwinkle Way, Sanibel; 472-2684.* Services Sun. at 8 a.m. (communion), 9 a.m. (contemporary), 10:30 a.m. (traditional). Sunday school Sun. at 9 a.m. General fellowship Sun. at 10 a.m. in courtyard. LOGOS youth program.
■ **Sanibel Congregational United Church of Christ.** *2050 Periwinkle Way, Sanibel; 472-0497.* Sun. services Nov.–April at 7:45 a.m., 9 a.m., 11 a.m.; May–Oct. at 7:45 a.m., 10 a.m.
■ **Unitarian-Universalist Society of the Islands.** *Services at the Congregational United Church of Christ, 2050 Periwinkle Way, Sanibel; 472-1646.* Services Sun. 7:30 p.m. Nov.–April except last Sun. of month.
■ **Unity of the Islands.** *Services at the Congregational United Church of Christ, 2050 Periwinkle Way, Sanibel; 278-1511.* Services first Sun. of month at 4 p.m.
■ **Vineyard Christian Fellowship of Sanibel.** *4115 Sanibel-Captiva Rd., Sanibel; 472-1018.* Interdenominational services Sun. 9 a.m. (time may vary).

Shipping

■ **Arundel's Hallmark Pack & Ship**
1626 Periwinkle Way, Sanibel; 472-0434
14900 Captiva Dr., Captiva; 395-0434
■ **Pak 'N' Ship**
2402 Palm Ridge Rd., Sanibel; 395-1220
■ **Quik Pack & Ship**
1713 Periwinkle Way, Sanibel; 472-0288

Federal Express drop boxes:
- ■ 1101 Periwinkle Way, Sanibel
- ■ 1506 Periwinkle Way, Sanibel
- ■ 1626 Periwinkle Way, Sanibel
- ■ 1713 Periwinkle Way, Sanibel
- ■ 2477 Periwinkle Way, Sanibel
- ■ 2402 Palm Ridge Rd., Sanibel
- ■ 695 Tarpon Bay Rd., Sanibel
- ■ 11500 Andy Rosse Ln., Captiva

Television

- ■ **ABC** WBR-TV, Cable 7
- ■ **CBS** WINK-TV, Cable 5
- ■ **FOX** WFTX-TV, Cable 4
- ■ **NBC** WBBH-TV, Cable 2

Weddings

You can buy a marriage license at the Lee County Office of the Clerk of the Circuit Court (335-2273). The cost is $88.50 cash (no personal checks accepted); the license is valid 60 days. Both you and your betrothed must appear at the court in person; you both need a valid photo ID (a driver's license is OK), and your Social Security number. Florida does not require a blood test. There is no waiting period unless you are a Florida resident; Floridians must wait three days after getting their license before they can marry. **Weddings may be performed** by a clergy member, Notary Public, member of the judiciary or Clerk of the Court.

Index